Encyclopaedia of Thermodynamics

Encyclopaedia of Thermodynamics

Volume 1

Thermodynamic Systems

Udai Veer

ANMOL PUBLICATIONS PVT. LTD.
NEW DELHI - 110 002 (INDIA)

ANMOL PUBLICATIONS PVT. LTD.
4374/4B, Ansari Road, Daryaganj
New Delhi - 110 002
Ph.: 23261597, 23278000
Visit us at: www.anmolpublications.com

Encyclopaedia of Thermodynamics

First Published, 2006

ISBN 81-261-2612-4 (Set)

PRINTED IN INDIA

Published by J.L. Kumar for Anmol Publications Pvt. Ltd., New Delhi - 110 002 and Printed at Mehra Offset Press, Delhi.

Contents

Preface

Science is the sermon of today. In every walk of life, since childhood till old age, science plays an important role in the life of man. Uptill some decades back, studied merely as a subject in schools and colleges, Science has assumed the status of an essential part of life today. With the expansion of literacy among the masses and growing awareness, they have woken up from the slumber of ignorance and have started taking up things more sincerely and thoughtfully. It is one of the reasons that our nation is heading towards new goals day after day.

It is a vast subject to study and talk about. For the purpose of ease, Science has been divided into more than two hundred and fifty branches. Specialisation in any branch can be obtained at different institutions and universities, scattered all over the country and abroad.

Apart from other super-specific branches, Physics, Chemistry, Zoology, Botany, Mathematics and Computronics, etc. are more general and closer to daily human life. From the point of view of the cut-throat competition in educational world, a thorough study of the above-mentioned branches, proves very handy, for the new aspirants, especially and for teachers and tutors, generally.

An urgent need was being felt for such a series of books, encompassing the commonly-studied and discussed branches of Science of equal significance and presented in an easy language, understood by all. With God's grace, our dream came true and we are happy, while introducing the dream-series for the lovers of knowledge.

The series consists of many a title, covering various branches of Science, such as Physics, Chemistry, Zoology, Botany, Geology, Agriculture Science, Computer Science, Mathematics and so on. The series is important and a must not only for students but also for teachers. Explained in simple language and aided with graphics and diagrams, everything, between the two ends—title and back cover—of the series, has been presented in a perfect way.

This set of books is devoted to **Thermodynamics**, which comprises two volumes: (1) Thermodynamic Systems and (2) Thermodynamic Laws.

It is hoped that the series would serve both, the students community and the academics, equally and will prove to be a boon for those, aspiring for newer heights. Nevertheless, any suggestions for further betterment of the endeavour will be highly appreciated and solicited.

Editor

1

Thermodynamic Systems

Thermodynamics is that branch of physics which is mainly concerned with the transformation of heat into mechanical work. It involves the study of the interaction of one body on another, the interaction being described in terms of the quantities heat and work. Thus, thermodynamics may be said to be an exact mathematical since that describes the inter-relationship between heat and mechanical energy. With the passage of time, however, the scope of thermodynamics has increased and now-a-days it coveres all the branches of physics as well as chemistry whenever we have to deal with heat and its relationship to energy in any of its diverse forms-electrical, mechanical, chemical, magnetic etc.

Thermodynamics constitutes a very subtle study in physics, chemistry and engineering and has innumerable applications. It is based upon two general laws of nature which govern the conversion of heat into work and vice versa. The first law represents the connection between heat and mechanical work while the second law depicts the manner in which these energy changes take place.

Thermodynamics takes no account of the atomic constitution of matter. Here we only deal with macroscopic (large scale) properties and do not go into the detailed structure. Thermodynamics is an empirical or experimental science and the laws of thermodynamics have been developed by means of observation and experiments over a long period of years. Thermodynamical formulae have necessarily the same general validity as the two laws from which they are derived.

A definite quantity of matter bounded by some closed surface is known as a system. A system might be the gas contained in a cylinder with a movable piston, or it might be the air inside a tyre that is being inflated. It may have an enclosure made by means of physical walls or imagined to be made by a mathematical surface, which serves as the boundary of the system and comprises everything of thermodynamic interest contained within the boundary. Thus, the boundary of the system may be actual or imaginary, it may have a fixed position or may be distorted as when the system is expanding. The existence of the boundary is, however, essential to visualize the system distinctly from rest of the universe. In an ordinary sense, a thermodynamic system may contain no substance at all, but may consist of radiant energy or an electric and magnetic field. Anything outside this system which can exchange energy with it and has a direct bearing on the behaviour of the system is called as its surroundings. The choice of the system and the surroundings is, however, quite arbitrary. What might be considered as part of the surroundings of a system by one person may be considered as part of the system itself by another person.

State of a System: It is defined by ascribing values to a sufficient number of variables. In mechanics, the state of a body is specified by means of a few measurable quantities such as the position of centre of mass, its velocity and acceleration etc. In thermodynamics, however, we disregard these mechanical coordinates and instead consider the interior of the system; find out directly or indirectly by experiments some measurable quantities upon which the internal state of the system depends. It is found by experiments that for all thermodynamic purposes, the state of a system can be represented by specifying its pressure P, volume V and the temperature T provided the system is in equilibrium. These parameters describing the state of the system are called *thermodynamic variables or thermodynamic coordinates.* Thus a *thermodynamic system is one which can be described in terms of thermodynamic coordinates.* General thermodynamic system are gases, vapour, steam, solid, liquid, surface films, stretched wires, thermocouples, magnetic materials, electrical cells etc.

Systems of Three Kind

1. *Open System :* Systems which can exchange both matter and energy with the surroundings are known as open system.
2. *Closed System :* Those systems which can exchange only energy (not matter) with the surroundings are called closed system.
3. *Isolated System:* If a system is not influenced in any way by its surroundings, it is said to be isolated. Thus an isolated system exchanges neither matter nor energy with its surroundings.

Thermal Equilibrium and Zeroth Law of Thermodynamics : A thermodynamic system is said to be in *thermal equilibrium* if all parts of it are at the same temperature and this temperature is the same as that of the surroundings. If these conditions are not satisfied, a change of state will take place until a thermal equilibrium is reached. A change of state here means a change in the thermodynamic state of a system, it does not mean at all a change from solid state to liquid state or gaseous state and other like changes. In thermodynamics, they are known as change of phase.

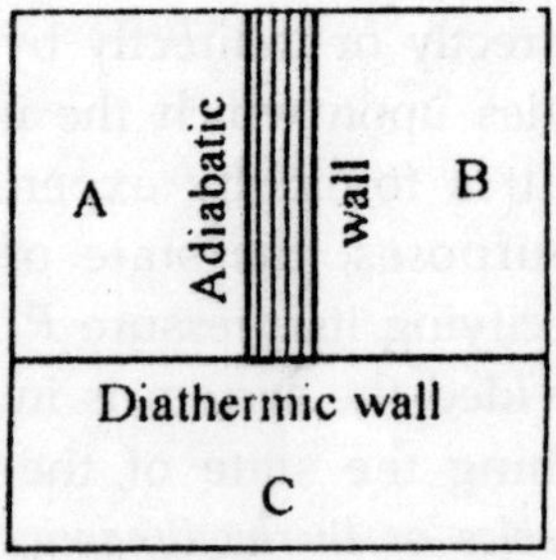

Now consider two systems *A* and *B* separated from each other by an adiabatic wall (one that conducts no heat), but each being in contact of a third system *C* through a diathermic wall which permits heat to pass (Fig. above). Then the system A will be in thermal equilibrium with system *C* and similarly system *B* will also be in thermal equilibrium with system C. Now if the adiabatic wall between systems *A* and *B* is replaced by a diathermic wall, experiments show that no further change occurs in systems *A* and *B* indicating thereby that system *A* was also in thermal equilibrium with system *B*. This important experimental fact leads us to the following general conclusion:

If two systems are in thermal equilibrium with a third system then they must be in thermal equilibrium with each other.

This statement is called the Zeroth law of thermodynamics.

Illustration : A simple illustration of zeroth law is provided in case of two gases *A* and *B* enclosed in vessels and a mercury thermometer C. If there is no change in the length of mercury thread on placing the thermometer *C* either in contact of gas vessel containing A or that of vessel containing *B*, then there will no change if A and B are brought in thermal contact with each other. This is in accordance of zeroth law.

Zeroth Law and Temperature

Zeroth law forms the basis of concept of temperature. All these three systems can be said to possess a property that ensures their being in thermal equilibrium with one another.

This property is known as *temperature.* We may, therefore, define the *temperature of a system as the property that determines whether or not the system is in thermal equilibrium with the neighbouring systems.* It is obvious that if two systems are not in thermal equilibrium, they will be at different temperatures.

The Equation of State

In order of describe the thermodynamic state of a system, certain variables are required. These variables, in general, are the mechanical variables supplemented by the thermal state of the system. Thus, for example, the thermodynamic state of a gaseous system can be specified by its pressure P, volume V and temperature T. But out of these three variables P, V and T, only two are independent variables and the third may be considered as their function. Thus, if we imagine a constant mass of the gas in a vessel and fix its pressure at some arbitrary value, then its volume can acquire a range of values for different values of temperatures, but if the temperature is also fixed, volume can have no arbitrary value i.e., the value of V at equilibrium is fixed.

Thus, the volume must be functionally related to the temperature and pressure, so that only two of the three are independent. If V and T are chosen arbitrarily by us, the value of P at equilibrium is determined by nature. In this way if we fix any two variables out of the three the third one is automatically fixed or in other words, only two parameters are necessary and sufficient to specify uniquely the state of a system. *The equation connecting the pressure, the volume and the absolute temperature of a substance is called the equation of state of that substance.* Thus for a gas forming the system, these three quantities are not independent and are connected by a relationship of the general form

$$f(P,V,T) = 0$$

which is called the equation of state.

For example, the equation of state for an ideal gas in a static condition is represented as

$$PV = RT$$

where R *is* the universal gas constant and its value is found to be 8.314 Joule/gm mole-K or 8.314×10^3 Joule/kg mole-K and V is the molar volume. For a Vander Waal's gas the equation of state is

$$\left(P+\frac{a}{V^2}\right)(V-b)=RT$$

where a and b are Vander Waal's constants.

For an equation of state, it should be noted:

1. Every thermodynamic system has its own equation of state independent of the others.
2. An equation of state expresses the peculiar behaviour of one individual system which distinguishes it from the others.
3. An equation of state is not a theoretical deduction from thermodynamics but is an experimental addition to it.
4. The equation of state is not applicable to systems not in thermodynamic equilibrium.

Thermodynamic Equilibrium

If an isolated system is left to itself, the pressure and temperature at a given point approach to a steady value with the passage of the time. When these quantities have attained a steady value, the system is said to be in a state of thermodynamic equilibrium. In such a state, the temperature, pressure and other thermodynamic variables of a homogeneous system are the same at all points. The state of a system in thermodynamical equilibrium can be represented by specifying its pressure, volume and temperature which are independent of the time. Thus, a system in thermal, mechanical and chemical equilibrium is said to be in thermodynamical equilibrium. In the state of thermodynamical equilibrium, the temperature and pressure are the same at all points and each phase is homogeneous.

Thus, a system in thermodynamic equilibrium must fulfil the following conditions :

(i) *Mechanical Equilibrium:* For a system to be in mechanical equilibrium, there should be no unbalanced forces existing between different parts of the system or between the system and the surroundings.

(ii) *Thermal Equilibrium:* For a system to be in thermal equilibrium, the temperatures at all parts of the system must be the same and should be identical with that of the surroundings.

(iii) *Chemical Equilibrium:* For a system to be in chemical equilibrium, the composition of the system should remain fixed and definite.

A system which satisfies all the above three equilibria, i.e., uniformity of pressure, temperature and chemical composition, is said to be the thermodynamic equilibrium.

Thermodynamic Processes

Any change which causes the thermodynamic variables of a system to have a new set of values is called a process. Whenever work is done by the surroundings or by some external agency or vice versa, a process is said to be occurring.

If a process can be carried out in such a way that the effects produced by it on the system as well as the surroundings can be completely restored to their initial states and no changes are left in any of the systems taking part in the process or in the surroundings, then the process is said to be *reversible process*.

In a reversible process the system passes through a succession of equilibrium states or states that differ from equilibrium only infinitesimally. Such a process must take place slowly in a controlled manner. If a process does not satisfy these conditions, it is called irreversible. Thus, the processes which produce a permanent change in the thermodynamic state of the system and cannot be retraced in the opposite order, are known as *irreversible processes.*

We shall, however, deal with these processes in more detail in chapter 2 (c.f. § 2.2).

External and Internal Works

Work is defined as the way by which a system may exchange energy with its surroundings. If a force $\vec{F}$ produces an infinitestimal displacement $\overrightarrow{dS}$, then the work done is defined in terms of the dot product of the two vectors, i e.,

$$dW = \vec{F} \cdot \overrightarrow{dS}$$

If a force is exerted by the system as a whole on its surroundings and a displacement takes place, the work that is done either by the system or on the system is called the *external work.* For example, a gas contained in a cylinder at a uniform pressure while expanding pushes the piston out and thus does external work on its surroundings. On the other hand, the work done by one part of the system on another part of the same system, is called the *internal work.* For example, in an actual gas there exist intermolecular attractions and when such a gas expands, the work is done against these mutual attractions between the molecules, which is the internal work. Thus, interactions of the molecules or electrons of a system on one another constitute internal work.

Let us now consider a constant mass of a gas confined in a piston and a cylinder at a constant pressure P. We may consider the gas as the system and the piston and cylinder as the surroundings. If we represent the area of the piston by *A*, the force exerted by the gas on the piston is the product of the pressure and the area, i.e.,

$$F = PA$$

because pressure is defined as force per unit area.

If now we allow the gas of the cylinder to expand and the piston moves out a distance *dS*, the external work done by the system,

$$dW = PA\, dS$$

But A $dS = dV$, the increase in the volume of the system. Hence, we have

$$dW = P\, dV$$

For a finite change in volume, from V_1 to V_2, the work done

$$W = \int_{V_1}^{V_2} P\, dV$$

Internal Energy of a System

Every thermodynamic system has within itself a definite quantity of energy. For example,

(a) a liquid freezes into a solid yielding heat energy

(b) zinc and copper sulphate in a Daniel cell react producing electrical energy,

(c) a mixture of hydrogen and oxygen gas explodes providing mechanical energy.

This store of energy which is not apparently shown by a system but is still capable of doing work, is called the *internal energy or intrinsic energy.*

Thermodynamics has nothing to do with the source of this energy, it is sufficient for its purpose that such an energy exists. However, according to kinetic theory of matter, matter is made up of ıarge number of tiny particles called molecules which are in a state of constant rapid motion and hence possess kinetic energy. The total kinetic energy of the molecules is called the *internal kinetic energy* while the potential energy arising due to the intermolecular attractions is called the *internal potential energy*. The energy of the system may be increased by causing its molecules to move faster (a gain in kinetic energy) or by causing the molecules to move against inter-molecular forces into a configuration of greater potential energy or both. The sum of the kinetic and potential energies of the molecules of a body, apart from any kinetic or potential energy of the body as a whole, is called its internal energy and is denoted by the letter U. *Thus internal energy may be defined as the energy possessed by a body or by a system due to molecular motion and molecular configuration.* The energy due to molecular motion is called *internal kinetic energy* U_k and that due to molecular configuration *internal potential energy* U_p, *i.e.*,

$$U = U_k + U_p$$

Let us now consider an isolated thermodynamic system subjected to external forces. Let an amount of work $W_{i \to f}$ be done on the system in going from a state i to a state f adiabatically and the internal energies in the initial and final states be represented by U_i and U_f respectively, then assuming the law of conservation of energy,

$$U_f - U_i = - W_{i \to f}$$

where the minus sign is introduced because the work is done on the system.

The physical interpretation of the difference $U_f - U_i$ is the increase in the energy of the system. Thus, there exists an energy function the difference in the two values of which is just equal to the work done in going from i to f. Thus,

(i) Internal energy is a function of state (thermodynamic coordinates).

(ii) Internal energy depends upon the initial and final states only and is *independent of* the *path followed, i.e.,*

$$\Delta U = U_f - U_i$$

(iii) Internal energy *depends upon* the *temperature.*

(iv) If a system returns to its initial state after passing through various states, the change in internal energy is zero, *i.e., in a cyclic* process.

$$\oint dU = 0$$

(v) Internal energy of an ideal gas is totally kinetic ($U_p = 0$). Thus

$$U = U_k = (3/2)\ \mu\ RT.$$

Concept of Heat

Heat may be defined as the *energy in transit.* If flows from one point to another. If the flow of heat stops, we can not use the word heat *i.e.,* the word heat may be used only when there is transfer of energy between two or more thermodynamic systems. When two bodies of different temperatures are placed in contact with each other, the two temperatures approach each other and after some time become equal. This phenomenon is described by saying that energy has flown from one body to another. Thermodynamically *this flow of energy from one body to another on account of difference of temperature between them is called heat.*

Thus, the flow of heat is a non-mechanical energy transfer brought about by a temperature difference between two bodies. For example, if a system loses energy by conduction or radiation, we may say that it is giving up heat as the process of conduction or radiation depends upon the temperature difference between the system and its surroundings. But if a system loses energy by other methods, generally by working against external mechanical forces, it is said to do work. Thus, work is the transfer of energy without a temperature difference. Hence, heat must be carefully distinguished from work.

Regarding heat, it is worth noting that :

(i) Heat ΔQ is path dependent.

(ii) It is taken positive if the system absorbs heat and negative if releases it.

(iii) In case of solids and liquids

$\Delta Q = \mu L \rightarrow$ when there is change of state

$\Delta Q = \mu C \Delta T \rightarrow$ when temperature changes by ΔT

(iv) For gases when heat is absorbed and temperature changes, for *m* gm mole of gas

$(\Delta Q)_v = \mu C_v \Delta T \rightarrow$ when volume is constant

$(\Delta Q)_p = \mu C_p \Delta T \rightarrow$ when pressure is constant

Indicator Diagram

It is just a graph between pressure and volume of a system undergoing an operation. It is used to find out the indicated horse power of an engine and provides us a mean to represent the working of an engine and to calculate the work obtained from it.

Suppose, a given mass of a working substance, say a gas, is contained in a cylinder fitted with a movable and frictionless piston. Let P and V be the pressure and volume of the gas and A the area of cross-section of the piston. Suppose now, the gas expands very slowly pushing the piston out through an infinitestimal distance dx. The external work done during expansion by the gas

$$dW = \text{Force exerted on the piston} \times \text{displacement}$$

$$= (P \times A) \times dx = P \times (A \times dx)$$

$$= P\,dV$$

where $dV = A \times dx$ is the increase in volume of the gas.

If the gas expands from an initial state (P_1, V_1) to a final state (P_2, V_2), then total work done by during the process

$$W = \int_{V_1}^{V_2} P\,dV$$

Let us now represent the initial state (P_1, V_1) of the gas by the point *A* and its final state (P_2, V_2) by the point *B* on a pressure volume diagram (Fig. below). The intermediate changes of the process are represented by different points on the line *AB* which, therefore, represents the whole operation.

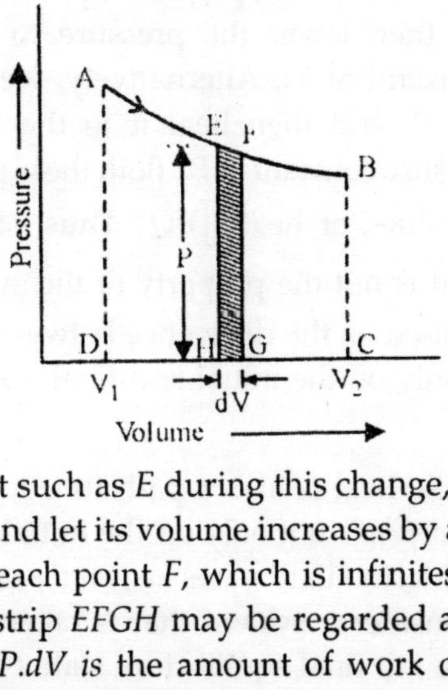

At any point such as *E* during this change, let the pressure of the gas be *P* and let its volume increases by an infinitestimal amount *dV* to reach point *F*, which is infinitestimally close to *E*. The shaded strip *EFGH* may be regarded as a rectangle of area *P.dV*. But *P.dV is* the amount of work done by the gas during the small expansion *dV*. Thus, the work done by the gas *P.dV is* represented by the area of the shaded strip.

The total work done by the gas in moving from *A* to *B* is clearly equal to the sum of areas of all such strips *i.e.*, equal to the area *ABCD* bounded between the curve *AB* and the volume axis from V_1 to V_2.

Hence, $$W = \int_{V_1}^{V_2} P.\,dV = \text{Area of } ABCDA$$

Thus, on an indicator diagram *the area between the P-V curve the volume axis and the two ordinates at the extremities of the P-V diagram gives directly the work done during the operation.*

Dependence of Heat and Work on Path

When a system changes from an initial state 1 (with coordinates P_1, V_1, T_1) to state 2 (with coordinates P_2, V_2, T_2) the *amount of heat transferred depends not only on the initial and final states but also on the intermediate states i.e.,* on the path of process. The heat flown into the system will be different for different paths. It depends on how the system is heated. For example, we can heat a gas at constant pressure P_1 until its temperature rises to T_2 and then lower the pressure to P_2 keeping the temperature constant at T_2. Alternatively, we can first lower the pressure to P_2 and then heat it to the temperature T_2, keeping the pressure constant at P_2. Both these processes would give different values of heat $\int \delta Q$. Thus, δQ *is* an inexact differential i.e., it is not the property of the system and $\int \delta Q$ cannot be expressed as the difference between two quantities which depend only on the initial and final states. Thus *heat is a path function.*

The work, like heat is also a path function. *Work done by* (or on) *a system depends* not only on the initial and final states but also on the intermediate states i.e., *on the path of the process.* For example, consider a system that is taken from an initial equilibrium state *i* to final equilibrium state *f* by two different paths *a* and *b* as shown in the indicator diagram (Fig. below). As is seen from the diagram the areas under the curves *ia f* and *ib f* are different and hence the *work done in taking the system from state i to state f will be different for different paths.* Thus, work W is not a property of the system and it can not be expressed as the difference between two quantities that depend entirely on the initial and final states. In other words the work *W is not a thermodynamic function* like pressure, temperature or internal energy.

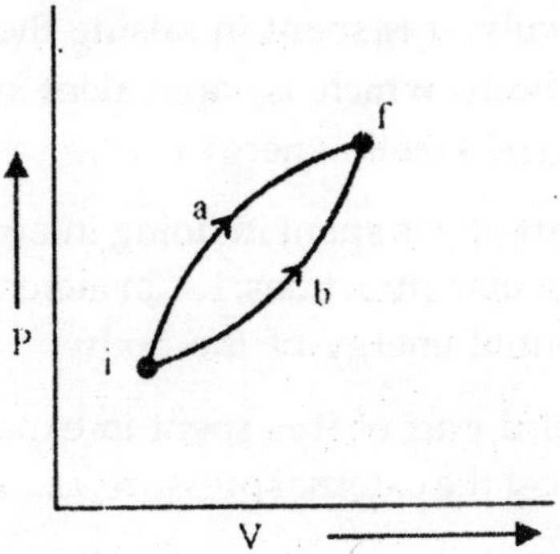

The First Law

The first law of thermodynamics may be stated in two slightly different forms

(1) The first form of this law is one which expresses the equivalence between mechanical work and heat energy. According to it, "*When mechanical work is spent in producing heat, a definite quantity of heat is produced for each unit of work spent; and conversely, when heat is employed to do work, the same definite quantity of heat disappears for every unit of work obtained.*" Thus if W is the amount of work, then amount of heat Q that can be produced by it, is given by

$$W = JQ$$

where $J = 4.2 \times 10^7$ ergs/cal = 4.2 Joules/cal. = 4.2×10^3 Joule/kilo cal. is a universal constant, called the mechanical equivalent of heat.

This form of law is true only if the whole of work done is used in producing heat or vice versa.

(2) The second statement of this law is a particular form of the general law of conservation of energy. Let us suppose that a quantity dQ of heat is supplied to a body. It is, in general, spent in three ways

(i) Partially, it is spent in raising the temperature of the body which is equivalent of increasing its internal kinetic energy.

(ii) A part of it is spent in doing internal work against molecular attractions, i.e., in increasing the internal potential energy of the body.

(iii) A third part of it is spent in expanding the body against the external pressure, *i.e.*, in doing external work.

If dU_k and dU_p are respectively the change in internal kinetic energy and internal potential energy and dW be the external work done then since "energy can neither be created nor destroyed but only converted from one form to another", we have

$$dQ = dU_k + dU_p + dW$$

But $dU_k + dU_p = dU$, the increase in total internal energy of the body.

Hence, $dQ = dU + dW$

or $\Delta Q = \Delta U + \Delta W$ ... (1)

This equation represents the *differential form of first law of thermodynamics,* which may therefore be stated. *"In all transformations, the energy due to heat units supplied must be balanced by the external work done plus the increase in internal energy."*

While using eq. (1), two things should be kept in mind :

(i) ΔQ. ΔU and ΔW are all to be measured in the same units i.e., all the three either in Joules or in calories or kilo calories.

(ii) If the heat is taken by the system then ΔQ is positive and if it is given by the system then ΔQ is negative. Similarly, if the work is done by the system then ΔW

is positive and if the work is done by some external agency on the system then ΔW is negative.

Significance of First Law

The first law of thermodynamics establishes an exact relationship between heat and work. According to it, a definite quantity of heat will produce a definite amount of work and vice versa. It denies that work or energy can be created out of nothing. It means that it is impossible to construct a thermal machine which may operate without any expenditure of fuel and may thus create energy out of nowhere. A machine that would do this could run itself and is generally called a perpetual motion machine of the first kind. The first law rules out such mechanics and is therefore sometimes defined in the form, *"Perpetual motion machine of the first kind is impossible"*. The fact that no exception to the first law has been found yet and no one has succeeded in building such a machine forms an experimental basis for the first law.

Three ideas are included in first law of thermodynamics:

(i) Heat is a form of energy in transit.

(ii) Conservation of energy takes place in thermodynamic systems.

(iii) Every system in equilibrium, possesses an internal energy which is a function of state, i.e., depends only upon the initial and final states of the system.

Specific Heats of a Gas

Materials differ from one another in the amount of heat required to produce a given change of temperature in a given mass. Suppose that an amount of heat Q is given to a body. If its temperature rises by ΔT then the *heat capacity* of the body is defined as

$$\text{Heat capacity} = \frac{Q}{\Delta T}$$

Heat capacity per unit mass is known as specific heat and represented by the letter c. Thus,

$$c = \frac{\text{Heat capacity}}{\text{mass}} = \frac{Q}{m.\Delta T} \qquad \text{... (1)}$$

Thus, specific heat of a material is the *quantity of heat required so as to raise the temperature of unit mass of the material through 1 degree.* The unit of specific heat in C.G.S. system is *calories per gm. per.* °C. It may also be expressed as *'Joule per kg. per °C'*. For example, Specific heat of water = 1 cal/gm°C

= 1 kilo calorie/kg °C

= 4.18 × 10^3 Joule/kg °C

The above definition of specific heat is sufficient for solids and liquids but not for gases because from eq. (1), specific heat of a gas may vary from zero to infinity. For example, if a gas, is compressed there is a rise in temperature without supplying any heat to the gas. Thus,

$$c = \frac{Q}{m.\Delta T} \quad 0 \text{ Because } Q = 0$$

On the other hand if heat is supplied to the gas and the gas is allowed to expand such that there is no rise in temperature, then

$$c = \frac{Q}{m \times 0} = \infty \text{ Because } \Delta T = 0$$

Hence, in order to find the value of the specific heat of a gas, either the pressure or the volume has to be kept constant. Consequently, we have two specific heats:

(i) The specific heat at constant volume which is defined as the amount of heat required to raise the temperature of unit mass of a gas through 1°C when its volume is kept constant. It is denoted by c_V and given by:

$$c_V = \left(\frac{\Delta Q}{\Delta T}\right)_V$$

(ii) The specific heat at constant pressure which is defined as the amount of heat required to raise the temperature of unit mass of the gas through 1°C when its pressure is kept constant. It is denoted by *cp* and given by

$$c_P = \left(\frac{\Delta Q}{\Delta T}\right)_P$$

Difference Between Two Specific Heats C_p and C_V of a Gas

c_p is greater than c_V: When a gas is heated at constant pressure, it expands. Therefore, heat supplied at constant pressure is partly used in doing work and partly in raising temperature of the gas. On the other hand, when gas is heated at constant volume, no work is done ($dW = P \,.\, dV = 0$) and hence whole of the heat supplied is used to raise the temperature of the gas. Thus, more heat is required for increasing the temperature of the gas through 1°C at constant pressure than at constant volume. Hence, specific heat of a gas at constant pressure is greater than the specific heat at constant volume, i.e., $c_p > c_V$.

In order to bring out clearly the application of the first law of thermodynamics, let us find a relation expressing the difference between two specific heats of a gas. For it, consider two isothermals *AB* and *CD* drawn for 1 mole of an ideal gas at close temperatures *T* and $(T + \Delta T)$. Let the initial state of the gas be represented by point *M* on lower isothermal at *T*. Let it now be heated at constant volume until its temperature rises to $T + \Delta T$ and its new pressure correspond to point *L* on the upper isothermal *CD*. As the volume of the gas is kept constant no external work has been done by the gas and hence the change in internal energy the process $M \rightarrow L$ is given by first law to be

$$U_L - U_M = \Delta Q - \Delta W = c_V \Delta T \qquad (\because \Delta W = 0) \quad \ldots (i)$$

Again starting from the same initial condition of the gas represented by *M*, let it now be heated at constant pressure until its temperature rises to $(T + \Delta T)$ and its new volume $(V + \Delta V)$ corresponds to point N on upper isothermal *CD*. Hence the change in internal energy in the process M $\rightarrow$ *N* is

$$U_N - U_M = \Delta Q - \Delta W = cp \Delta T - P \Delta V \quad \ldots (ii)$$

because in this case $\Delta Q = c_p \Delta T$ and work performed by the gas $\Delta W = P. \Delta V$.

Since, the gas is supposed to be a perfect one, the perfect gas equation $(PV = RT)$ holds good. In the initial position at *M*, the pressure, volume and temperature of the gas were respectively *P*, *V* and *T* while at *N* volume becomes $(V + AV)$ and temperature $(T + \Delta T)$. Hence, according to gas equation at M and *N* respectively, we have

$$PV = RT \quad \ldots (iii)$$

and $$P(V + \Delta V) = R(T + \Delta T) \quad \ldots (iv)$$

Subtracting eq. (iii) from (iv), we have

$$P. \Delta V = R. \Delta T$$

Putting this value of P. ΔV in eqn. (ii), we get

$$U_N - U_M = cp. \Delta T - R. \Delta T \quad \ldots (v)$$

As the *internal energy of a gas depends only upon its temperature* the same change in internal will take place in both the processes $M \rightarrow L$ and $M \rightarrow N$ because temperature changes from T to $T + \Delta T$ in both the processes. Thus

$$U_L - U_M = U_N - U_M$$

or $$c_V \Delta T = cp \Delta T - R. \Delta T \quad \text{[using eqn. (i) and (v)]}$$

or Dividing throughout by ΔT, we have

$$c_V = c_p - R$$

or $$c_p - c_v = R \quad \ldots (vi)$$

This formula is known as *Mayer's* formula. All these quantities c_p, c_v and *R* in this equation should be expressed in the same units either in Joule/mole'C or in cal/mol °C.

PROBLEMS

1. ***Calculate the change in internal energy when 0.004 kg. of air is heated from 0°C to 2°C. The specific heat of air at constant volume being 0.172 kilo cal /kg °C.***

Solution: From first law of thermodynamics

$$dQ = dU + dW$$

where all quantities are measured in same units, say kilo cal.

Here, air is heated at constant volume and hence no external work is done i.e.,

$$dW = p \,.\, dV = 0.$$

Let *m* be the mass, c_v the specific heat at constant volume and *dT* the rise in temperature of the air, the heat taken in by it is

$$dQ = m \times c_V \times dT \text{ kilo cal.}$$

Then from first law of thermodynamics

$$dU = dQ - dW = dQ$$

$$= m \times c_V \times dT$$

$$= 0.004 \times 0.172 \times 2$$

$$= 1.376 \times 10^{-3} \text{ kilo cal.}$$

2. ***What is the change in the internal energy per gm. of air if 15 gm. of air is heated from 0°C to 5°C at constant volume by adding 150 Gals. of heat ?***

Solution: The change in internal energy of the air is given by

$$dU = dQ - dW$$

Here, air is heated at constant volume. Hence

$$dW = P \,.\, dV = 0$$

$$\therefore \quad dU = dQ$$

$$= 150 \text{ cals.}$$

Therefore, change of internal energy per gm. of air is

$$\frac{dU}{15} = 10 \text{ cal./gm.}$$

3. *One gram of water becomes 1671 c.c. of steam when boiled at a pressure of one atmosphere. The heat of vaporisation at this temperature is 540 cal per gm. Compute the external work and the increase in internal energy.*

Solution: The external work done at pressure P for a change in volume of dV is given by

$$dW = P.\, dV.$$

Here $\quad P = 1$ atoms

$$= 0.76 \times 13.6 \times 10^3 \times 9.8$$

$$= 1.013 \times 10^5 \text{ N/m}^2$$

and $\quad dV = 1671 - 1 = 1670 \text{ c.c.} = 1670 \times 10^{-6} \text{ m}^3$

$\therefore$ External work

$$dW = 1.013 \times 10^5 \text{ N/m}^2 \times 1670 \times 10^{-6} \text{ m}^3$$

$$= 169.17 \text{ Joules}$$

$$= \frac{169.17}{4.18} \text{ cals.}$$

$$= 40.47 \text{ cals.}$$

Now, from the first law of thermodynamics

$$dQ = dU + dW.$$

Hence, change in internal energy

$$dU = dQ - dW.$$

HeredQ = 540 cals. and dW = 40.47 cals.

$$\therefore \qquad dU = 540 - 40.3$$

$$= 499.53 \text{ cals.}$$

4. At normal temperature (0°C) and normal pressure (1.013 × 10^5 N/m^2), when 1 gm. of water freezes, its volume increases by 0.091 cm^3. Calculate the change in internal energy.

[Given : Latent heat of ice = 80 cal /gm]

Solution: Heat given by 1 gm water when it freezes into ice

$$Q = mL = 1 \text{ gm} \times 80 \text{ cal/gm}$$

$$= 80 \text{ cal.}$$

$$= 80 \times 4.2 \text{ Joule}$$

$$= 336 \text{ Joule}$$

The external work donc against the atmosphere when its volume increases by 0.0091 cm^3

$$W = P \,.\, dV = 1.013 \times 10^5 \frac{\text{N}}{\text{m}^2} \times 0.091 \times 10^{-6} \text{ m}^3$$

$$= 0.0093 \text{ Joule}$$

By the first law of thermodynamics, the change in internal energy $\Delta L = Q - W = (-336 - 0.0093)$ Joule

$$= -336.0093 \text{ Joule}$$

Here Q is taken negative because heat is given by the system. Thus, internal energy decreases by 336.0093 Joule.

Quasi-static Process

A quasi-static process is defined as the process in which the deviation from thermodynamic equilibrium is infinitesimal and all the states through which the system passes during a quasi-static process may be treated as equilibrium states. Thus it may be defined as a *succession of equilibrium states*. For example, consider a system, a gas contained in a cylinder fitted with a movable piston. If the piston is pushed in very rapidly, the gas immediately behind the piston acquires kinetic energy and is set into turbulent motion which is a non-equilibrium state.

In such a case the work done on the system is not quasi-static and is not governed by the equation $dW = P\ dV$. If, however, the piston is pushed in at a infinitely slow rate, the system is at all times in a quiescent equilibrium state and then the process can be considered quasi-static. Thus during a quasi-static process, the system is at every moment infinitestimally near a state of thermodynamic equilibrium and all the states through which the system passes can be described by means of thermodynamic variables referred to the system as a whole. The vanishingly slowness of the process is an essential feature of a quasi-static process.

A quasi-static process is an idealized concept and the conditions for it can never be rigorously satisfied in practice. It is quite distinct from a real physical process because a real process always involves non-equilibrium intermediate states. A quasi-static process in contrast to a real process, does not involve considerations of rates, velocities or times. A real process is a temporal succession of equilibrium and non-equilibrium states whereas a quasi-static process is an ordered succession of equilibrium states.

Isothermal Process

If a thermodynamic system is perfectly conducting to the surroundings, and undergoes a physical process in such a way that its *temperature remains constant throughout*, the process is said to be *isothermal process*. In such a process, the heat developed or heat lost is given out to or taken from the surroundings in such a way that the temperature of the system remains constant.

If a gas contained in a metallic cylinder fitted with a piston is compressed slowly, the heat generated by the external work done upon the gas, conducts away to the surroundings through the walls of the cylinder. Similarly, if the gas is allowed to expand slowly, a slight cooling will be produced as the gas does work against the piston but an equivalent amount of heat will enter the system from the surroundings and equalize the temperature. Thus, in both cases the temperature remains constant and hence the processes are isothermal. The *melting of ice at its melting point and vaporization of water at its boiling point are the examples of* isothermal process.

The curves showing the variation in the volume of a substance when the pressure acting on it changes under isothermal conditions are called Isothermals.

The equation connecting the pressure P and volume V of one mole of a gas for an isothermal process is

$$PV = RT = \text{constant}$$

For *n* moles of a gas

$$PV = nRT$$

where $R = 8.3$ Joules/gm. mole $K = 8.3 \times 10^3$ Joules/kg mole K is the universal gas constant and T is the absolute temperature.

In isothermal changes, there is no change in temperature, no change in internal energy *i.e.* $dU = 0$

$$\therefore \qquad dQ = 0 + dW$$

$$\text{or} \qquad dQ = dW$$

Thus during an isothermal process,

Heat added (or subtracted) = work done by (or on) the gas.

Adiabatic Process

If a thermodynamic system is perfectly insulated from the surroundings and undergoes a process in such a way that *no exchange of heat takes place* between it and the surroundings, the process is said to be *adiabatic process.* In such a process no heat is allowed either to enter the system or to leave it, but all along the process, there is a change in temperature. If a gas is suddenly compressed, the external work done upon it, increases its internal energy and consequently its temperature rises. Similarly, if the gas suddenly expands, it does work against the surroundings at the cost of its internal energy and therefore, its temperature falls.

Thus, there is no exchange of heat between the gas and surroundings, being perfectly insulated from each other, and hence the processes are adiabatic. The compressions and rarefactions produced in air during the *propagation of sound waves* is also an example of adiabatic process. The sudden expansion of the enclosed air into the atmosphere when a *motor tyre bursts,* is approximately adiabatic and hence the tyre is cooled.

Thus, all processes that take place quickly or suddenly are adiabatic processes.

The curves showing the variation in the volume of a substance when the pressure acting on it changes under adiabatic conditions are called *Adiabatics.*

During the adiabatic process, no heat enters or leaves the system, but temperature changes, because work is done at the cost of internal energy.

For adiabatic change in a perfect gas, $dQ = 0$ and hence

$$0 = dU + dW$$

$$dU = -dW$$

i.e., change in internal energy = mechanical work done.

Thus, in the adiabatic process if the work is *done on the system (i.e. dW* is negative) then

$$dU = -(-dW) = dW$$

that is the internal energy of the system will increase by an amount equal to the work done on it.

On the other hand if the *work is done by the system (i.e.,* dW is positive), then

$$dU = -(+dW) = -dW$$

that is, the internal energy of the system will decrease by an amount equal to the work done by it.

Isobaric Process

If the working substance is taken in an expanding chamber in which the *pressure is kept constant,* the process is called *isobaric process.* In these processes the gas either expands or contracts to maintain a constant pressure and hence a net amount of work is done by the system or on the system. The amount of heat dQ given to the working substance is partly used in increasing the temperature dT of the working substance and partly used in doing external work. Considering unit mass of the working substance

$$dQ = 1 \times C_V\, dT + P.\, dV.$$

Thus, in isobaric process also, change in internal energy

$$dU = C_V \,.\, dT$$

Isochoric Process

If a substance undergoes a process in which the *volume remains unchanged,* the process is called an *isochoric process.* The increase of pressure and temperature produced by the heat supplied to a working substance contained in a non-expanding chamber is an example of isochoric process. Since, there is no change in volume, the work done in such a process is zero ($dW = P \,.\, dV = 0$) and, therefore, from first law

$$dQ = dU$$

Thus, whole of the heat supplied increases the internal energy of the system.

The heat supplied in such a process

$$dQ = C_V \, dT$$

Hence $\quad C_V \, dT = dU$

PROBLEMS

1. For each of the following cases of a process involving a closed system, calculate and write down the missing information. Q is positive when heat is supplied to a system and W is positive when work is done by the system. U_i and U_f refer to the internal energy of the system in the initial and the final states. All these quantities are expressed in Joules.

Process	Q	W	U_i	U_f	$\Delta U = U_f - U_i$
I	25	– 10		– 10	
II	– 10			65	-20
III		-20	80		20

Out of these processes, which one is adiabatic ?

Solution: From first law of thermodynamics

$$\Delta Q = \Delta U + \Delta W$$

$$Q = U_f - U_i + W \qquad (\because \ \Delta U = U_f - U_i)$$

Process I. $Q = 25$ Joules, $U_f = -10$ Joules, $W = -10$ Joules

$$\therefore \qquad 25 = -10 - U_i - 10$$

$$U_i = -45 \text{ Joules}$$

Hence $\quad \Delta U = U_f - U_i$

$$= -10 - (-45) = 35 \text{ Joules}$$

Process II. $\quad Q = -10$ Joules, $U_f = 65$ Joules,

$$\Delta U = -20 \text{ Joules.}$$

$$\therefore \qquad -20 = 65 - U_i$$

or $U_i = 85$ Joules

Also $W = Q - U_f + U_i$

$= -10 - 65 + 85 = 10$ Joules

Process III. $W = -20$ Joules, $U_i = 80$ Joules,

$\Delta U = 20$ Joules

$Uf = \Delta U + Ui$

$\therefore$ $= 20 + 80 = 100$ Joules

And $Q = U_f - U_i + W$

$= 100 - 80 - 20 = 0$ Joules.

Hence *Process III.* (For which $Q = 0$) *is adiabatic.*

2. *When a system is taken from a state i to a state f, in fig. below along the path iaf, we find dQ = 50 cals. and W = 20 cals. while along the path ibf, dQ = 36 cals. If* $U_i = 10$ ***cals.***

(i) ***What is*** U_f ***?***

(ii) ***What is W along the path ibf ?***

(iii) ***If W = – 13 cals. along the curved path f i, what is dQ for this path ?***

(iv) ***If*** $U_b = 22$ ***cals. what is dQ for the process ib ? for bf ?***

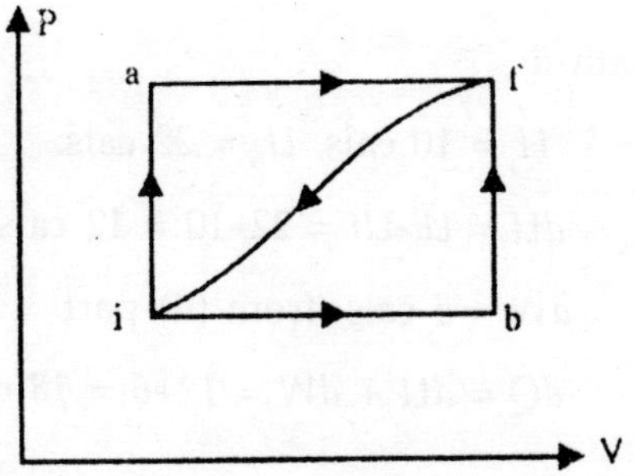

Solution : Given that $U_i = 10$ cals, $Q = 50$ cals, $W = 20$ cals. for the path *iaf.*

(i) From first law

$$Q = U_f - U_i + W$$

$$50 = U_f - 10 + 20$$

$$U_f = 50 + 10 - 20 = 40 \text{ cals.}$$

(ii) For the path *ibf*

$$U_i = 10 \text{ cals}, U_f = 40 \text{ cals.}$$

$$Q = 36 \text{ cals.}$$

$$W = Q - U_f + U_i = 36 - 40 + 10 = 6 \text{ cals.}$$

(iii) For the path *fi*

$$W = -13 \text{ cals.}$$

and $$dU = U_i - U_f = 10 - 40 = -30 \text{ cals.}$$

$$dQ = dU + dW = -30 - 13 = -43 \text{ cals.}$$

(iv) For the path *bf*

$$U_f = 40 \text{ cals}, U_b = 22 \text{ cals}$$

$$dU = U_f - U_b = 40 - 22 = 18 \text{ cals.}$$

and $$dQ = dU + dW = 18 + 0 = 18 \text{ cals.}$$

Because the work done ($dW = P \,.\, dV$) from b to f *is* zero as in the process volume remains constant.

For the path *ib*

$$U_i = 10 \text{ cals}, U_b = 22 \text{ cals.}$$

$\therefore$ $$dU = U_b - Ui = 22 - 10 = 12 \text{ cals.}$$

and $$dW = 6 \text{ cals. from (ii) part}$$

$\therefore$ $$dQ = dU + dW = 12 + 6 = 18 \text{ cals.}$$

2

Adiabatic Equation

Let us consider one gm. molecule of a perfect gas thermally insulated from its surroundings. Let it suffer a very small adiabatic expansion doing external work at the cost of its own internal energy. The temperature of the gas, therefore, falls.

If the volume of the gas increases by an infinitesimal amount dV against an external pressure P, the external work done by the gas in its expansion will be

$$dW = P \, . \, dV \quad \ldots \text{(i)}$$

Since, in a perfect gas, the molecules do not attract one another, the internal energy depends only on its temperature. Hence, the decrease in internal energy of the gas, suffering a fall dT in its temperature, is equal to the heat drawn from it i.e.,

$$dU = 1 \times C_v \times dT \quad \ldots \text{(ii)}$$

where C_V is the specific heat for one gm. mole of the gas at constant volume.

Applying now the first law of thermodynamics

$$dQ = dU + dW$$

or
$$dQ = Cv\ dT + P\ .\ dV$$

where dQ, $C_V\ dT$ and $P.\ dV$ are all measured in the same units i.e., either Joules or in calories.

Since, in an adiabatic change, no heat is supplied from outside,

Hence, $dQ = 0$.

Therefore, we have

$$C_V\ dT + P.\ dV = 0 \qquad \text{... (iii)}$$

Also for one gm. molecule of a perfect gas we have the relation

$$PV = RT \qquad (R \text{ being gas constant})$$

which on differentiation yields

$$P.dV + V.dP = R\ dT$$

or
$$dT = \frac{P.dV + V.dP}{R}$$

Putting this value in equation (iii),

$$C_v\left(\frac{P.dV + V.dP}{R}\right) + P.dV = 0$$

or
$$C_V\ (P.dV + V.dP) + R\ (P.dV) = 0$$

But from Mayer's relation

$$Cp - C_V = R$$

$$C_V\ (P.dV + V.dP) + (Cp - C_V)\ P.dV = 0$$

$$C_V.\ V.dP + Cp.P.dV = 0$$

Dividing by $C_V PV$,

$$\frac{dP}{P} + \frac{C_p}{C_V}\frac{dV}{V} = 0$$

But $\frac{C_p}{C_V} = \gamma$, the ratio of the principal specific heats of the gas.

Hence $$\frac{dP}{P} + \gamma\frac{dV}{V} = 0$$

Integrating, $$\log P + \gamma \log V = \text{constant}$$

or $$\log PV^{\gamma} = \text{constant}$$

or $$\boxed{PV^{\gamma} = \text{constant}} \quad \ldots \text{(iv)}$$

This is the equation connecting pressure and volume during adiabatic process and is known as *Poisson's law.*

There are two other forms of the above relation.

Relation between Temperature and Volume : Putting $P = \frac{RT}{V}$ in eq. (iv), we get

$$\frac{RT}{V}V^{\gamma} = \text{constant}$$

or $$\boxed{TV^{\gamma-1} = \text{constant}} \quad \ldots \text{(v)}$$

Relation between Temperature and Pressure : Putting $V = \frac{RT}{P}$ in eq. (iv), we get

$$P\left(\frac{RT}{P}\right)^{\gamma} = \text{constant}$$

or $$\frac{T^{\gamma}}{p^{\gamma-1}} = \text{constant}$$

or $$\boxed{TP^{(1-\gamma)/\gamma} = \text{constant}} \quad \ldots \text{(vi)}$$

Work Done During an Adiabatic Process : Let us consider one gm. mole of a perfect gas enclosed in a cylinder having perfectly insulated walls. Let it expand adiabatically from an initial volume V_1 to a final volume V_2. Then work done by the gas during this expansion is given by

$$W = \int_{V_1}^{V_2} P.dV$$

where P is the instantaneous pressure of the gas while suffering an infinitesimal expansion dV.

Now during an adiabatic process

$$PV^{\gamma} = K \text{ constant}$$

$$P = \frac{K}{V^{\gamma}}$$

Therefore, $$W = \int_{V_1}^{V_2} K\frac{dV}{V^{\gamma}} = K\int_{V_1}^{V_2} V^{-\gamma}\, dV$$

or $$W = K\left[\frac{V^{-\gamma+1}}{-\gamma+1}\right]_{V_1}^{V_2} = \frac{K}{1-y}\left[\frac{1}{V_2^{\gamma-1}} - \frac{1}{V_1^{\gamma-1}}\right] \quad \text{... (i)}$$

But in adiabatic process

$$P_1 V_1^{\gamma} = P_2 Va_2^{\gamma} = K$$

$$\therefore \quad w = \frac{1}{1-\gamma}\left[\frac{P_2 V_2^{\gamma}}{V_2^{\gamma-1}} - \frac{P_1 V_1^{\gamma}}{V_1^{\gamma-1}}\right]$$

$$= \frac{1}{1-\gamma}[P_2V_2 - P_1V_1] \quad \text{... (ii)}$$

$$= \frac{1}{1-\gamma}[RT_2 - RT_1]$$

$$= \frac{R}{\gamma-1}[T_1 - T_2] \quad \text{...(iii)}$$

Hence, the work done in adiabatic process depends only upon the initial and final temperatures T_1 and T_2. Thus, the *work done along any adiabatic between two isothermals is independent of the particular adiabatic.*

Work Done During an Isothermal Process : Work is done by a gas when it is allowed to expand isothermally. Let the initial and final volumes of 1 gm. mole of an ideal gas be V_1 and V_2 respectively. Then the work done for a very small change in volume dV

$$dW = P.dV$$

When the gas expands from volume V_1 to V_2, at constant temperature T, then work done

$$W = \int_{V_1}^{V_2} P\,dV$$

Since expansion is isothermal,

$$PV = RT \quad \text{or} \quad P = \frac{RT}{V}$$

$$\therefore \qquad W = RT\int_{V_1}^{V_2} \frac{dV}{V} = RT\log_e \frac{V_2}{V_1}$$

$$= 2.3026\ RT\log_{10} \frac{V_2}{V_1}$$

Work Done During an Isobaric Process : Let the initial and final volumes of one gm. mole, of a perfect gas enclosed in an expanding chamber in which the *pressure is kept constant* be V_1 and V_2. The work done for a very small change in volume dV

$$dW = P.\,dV.$$

When the gas expands from V_1 to V_2,

$$\text{Work done} \qquad W = \int_{V_1}^{V_2} P.\,dV$$

$$= P\,(V_2 - V_1).$$

because the pressure is kept constant in an isobaric process.

Laws of Expansion and Compression of a Gas : The general law giving relation between pressure and volume for expansion or compression of a gas can be conveniently written as

$$PV^n = \text{constant}$$

where n *is* a constant depending upon the nature of the gas. It can have values between 0 and ∞ but here we will take certain typical values.

(i) $n = 0$ for a perfect gas expanding at constant pressure.

(ii) $n = 1$ for a perfect gas expanding isothermally *i.e.*, at constant temperature.

(iii) $n = \frac{C_p}{C_v} = \gamma$ for a perfect gas expanding adiabatically *i.e.*, without taking or leaving any heat.

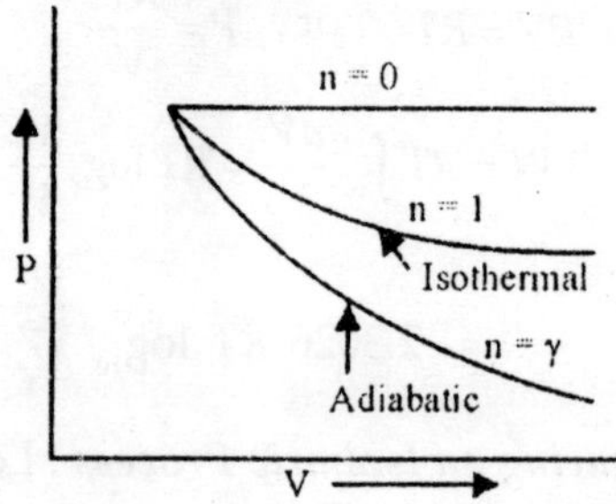

Greater is the value of n steeper is the curve for expansion or compression. Expansion curve for $n = 0$, $n = 1$ and $n = \gamma$ are depicted qualitatively on a $P - V$ diagram (Fig. above). $n = 0$ corresponds to expansion at constant pressure (isobaric process). $n = 1$ corresponds to an isothermal expansion and is given by PV = constant. $n = \gamma$ corresponds to adiabatic expansion and is given by PV^{γ} = constant.

Slopes of Adiabatic and Isothermals

An *isothermal curve* or simply *isothermal is* a graphical picture of isothermal change on $P - V$ diagram. Similarly, an *adiabatic curve* or *adiabatic* is a graphical representation of adiabatic change on a $P - V$ diagram.

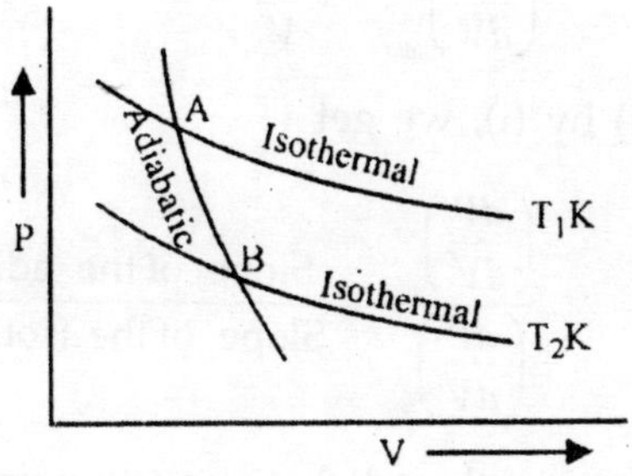

Let us draw two isothermals for a given mass of a perfect gas at constant absolute temperature T_1 and T_2 (Fig. above). Suppose thegas undergoes an adiabatic expansion starting from the pointA on the isothermal corresponding to temperature T_1. Now a gas undergoing adiabatic expansion does external work at the cost of its own internal energy and hence falls in temperature. At the end of adiabatic expansion it, therefore, seeks lower isothermal corresponding to lower temperature T_2. Thus, an adiabatic cuts a number of isothermals whereas two isothermals for the same gas cannot cut each other. It clearly indicates that *the slope of the adiabatic at a point is greater than that of the isothermal passing through the same point.* Analytically, it can be shown as follows:

For an isothermal process

$$PV = \text{constant}$$

Differentiating $P\,dV + V\,dP = 0$

or
$$\left(\frac{dP}{dV}\right)_{iso} = -\frac{P}{V} \qquad \text{...(i)}$$

For an adiabatic process

$$PV^{\gamma} = \text{constant}$$

Differentiating $P\gamma V^{\gamma-1} dV + V^{\gamma} dP = 0$

or $$\left(\frac{dP}{dV}\right)_{adia} = -\frac{\gamma P}{V} \qquad \text{...(ii)}$$

Dividing (ii) by (i), we get

$$\frac{\left(\frac{dP}{dV}\right)_{adia}}{\left(\frac{dP}{dV}\right)_{iso}} = \frac{\text{Slope of the adiabatic}}{\text{Slope of the isothermal}} = \gamma$$

Thus, the slope of the adiabatic at any point is γ times the slope of the isothermal at the same point or in other words, *the adiabatic curve is steeper than the isothermal curve.*

PROBLEMS

1. ***Prove that the work done by a perfect gas during a quasi static adiabatic expansion is given by***

(i) $$W = \frac{P_i V_i}{\gamma - 1}\left[1 - \left(\frac{P_f}{P_i}\right)^{(\gamma-1)/\gamma}\right]$$

(ii) $W = C_V [T_j - T_f]$

Solution: (i) From eq. (ii) of section 1.21, we have

$$W = \frac{1}{1-\gamma}[P_f V_f - P_i V_i]$$

$$= \frac{P_i V_i}{1-\gamma}\left[\frac{P_f V_f}{P_i V_i} - 1\right]$$

$$= \frac{P_i V_i}{1-\gamma}\left[1 - \left(\frac{P_f}{P_i}\right)\left(\frac{P_f}{V_i}\right)\right] \qquad \text{... (1)}$$

But from Poisson's law

$$P_i V_i^{\gamma} = P_f V_f^{\gamma}$$

$$\left(\frac{V_f}{V_i}\right)^{\gamma} = \frac{P_i}{P_f}$$

or
$$\frac{V_f}{V_i} = \left(\frac{P_i}{P_f}\right)^{\frac{1}{\gamma}} = \left(\frac{P_f}{P_i}\right)^{-\frac{1}{\gamma}} \quad \text{... (2)}$$

Hence, eq. (i), using (2), becomes

$$W = \frac{P_i V_i}{\gamma - 1}\left[1 - \left(\frac{P_i}{P_f}\right)\left(\frac{P_f}{P_i}\right)^{-\frac{1}{\gamma}}\right]$$

$$= \frac{P_i V_i}{\gamma - 1}\left[1 - \left(\frac{P_i}{P_f}\right)^{\left(1-\frac{1}{\gamma}\right)}\right]$$

or
$$\boxed{W = \frac{P_i V_i}{\gamma - 1}\left[1 - \left(\frac{P_f}{P_i}\right)^{\left(\frac{\gamma-1}{\gamma}\right)}\right]}$$
Proved.

(ii) From eq. (iii) of sec. 1.21, we have

$$W = \frac{R}{\gamma - 1}[T_i - T_f]$$

But $R = Cp - C_V$ and $\gamma = \frac{Cp}{C_V}$

$$\therefore \quad W = \frac{C_p - C_V}{\left(\frac{C_p}{C_V}\right) - 1}[T_i - T_f]$$

$$= \frac{C_p - C_V}{\left(\frac{C_p - C_V}{C_V}\right)}[T_i - T_f]$$

$$\therefore \quad \boxed{W = C_v\ [T_i - T_f]}$$
Proved.

2. *Calculate the work done when a gram molecule of a gas expands isothermally at 27°C to double its original volume. Given R = 8.3 Joule degree $^{-1}$ mole $^{-1}$.*

Solution: Work done during an isothermal process

$$W = 2.3026\ RT \log_{10} \frac{V_2}{V_1}$$

Here $T = 27^\circ C = 300\ K,\ V_1 = V,\ V_2 = 2V$

$$W = 2.3026 \times 8.3 \times 300 \log_{10} 2$$

$$= 2.3026 \times 8.3 \times 300 \times 0.3010$$

$$= 1725.8 \text{ Joule.}$$

3. *A mass of 10 gm. of hydrogen at 0°C is heated to 50°C at constant pressure and is then cooled to 0°C at constant volume. Compare the amounts of heat absorbed with that which would have been absorbed if the change from the initial state has been made isothermally.*

Given, C_p for hydrogen = 2.4 cal. /gm. mole/ K ***and*** R ***for hydrogen = 2.0 cal. /gm. mole /*** K.

Solution: Amount of heat absorbed when 10 gm. of hydrogen is heated from 0° to 50°C at constant pressure (curve *AB*, Fig. below) = $m\ cp\ 50$.

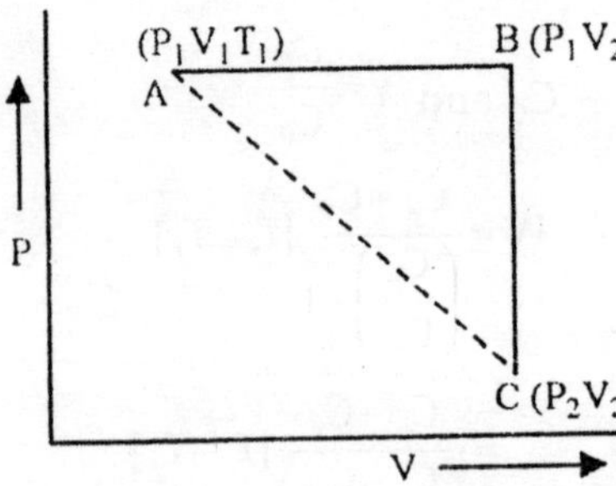

Amount of heat given out when 10 gm. of hydrogen is cooled from 50°C to 0°C along curve *BC* (isochoric process)

$$= m\ cv\ 50$$

∴ Net Absorption along *ABC*

$$= m \times 50\ (c_p - c_v) = m \times 50 \times r$$

$$= 10 \times 50 \times \frac{R}{2} = 500 \text{ calories} \qquad \ldots\text{(i)}$$

$$\text{(where } r = \frac{R}{2} \text{ is const. for 1 gm. hydrogen)}$$

Now amount of work done from A to C (isothermal process)

$$W = P_1V_1 \log_e\left(\frac{V_2}{V_1}\right) = 2.3026\, RT_1 \log_{10}\left(\frac{V_2}{V_1}\right)$$

But for AB,

$$\frac{V_1}{V_2} = \frac{T_1}{T_2}$$

$$\frac{V_1}{V_1} = \frac{T_2}{T_1} = \frac{273+50}{273} = \frac{323}{273}$$

$$\therefore \qquad W = \frac{10 \times 2 \times 273}{2} \times 2.3026 \log_{10} \frac{323}{273} \text{ cals.}$$

$$= 459.8 \text{ cals.} \qquad \ldots\text{(ii)}$$

Comparing eqn. (i) and (ii), the ratio of heat absorbed is 500: 459.8

4. ***A certain mass of a gas at 0°C is expanded to three times its volume under adiabatic conditions. Calculate the resulting temperature and pressure. Given*** γ ***for gas = 1.40.***

Solution: (i) $T_1V_1^{\gamma-1} = T_2V_2^{\gamma-1}$

Here $\qquad V_1 = V, V_2 = 2V, T_1 = 273$ K, $T_2 = ?$

$$\therefore \qquad T_2 273 = \left[\frac{1}{3}\right]^{1.4-1} = 176\text{K}$$

$$= -97°\text{C}$$

(ii) $\qquad P_1V_1^{\gamma} = P_2V_2^{\gamma}$

Here $V_1 = V, V_2 = 3V, P_1 = 1$ atmosphere, $P_2 = ?$

$$\therefore \quad P_2 = P_1\left[\frac{V_1}{V_2}\right]^\gamma$$

$$= 1\left[\frac{1}{3}\right]^{1.4} = 0.215 \text{ Atmosphere.}$$

5. ***A definite mass of a perfect gas is compressed adiabatically to half its original volume. Determine the resultant pressure if the initial pressure was one atmosphere. [y = 1.4 and ($2^{1.4}$ = 2.64)].***

Solution: Here P_1= 1 atmosphere,

$$P_2 = ?, V_1 = V, V_2 = \frac{V}{2}$$

and $\gamma = 1.4.$

For an adiabatic change

$$P_1 V_1^\gamma = P_2 V_2^\gamma$$

$$\therefore \quad P_2 = P_1\left[\frac{V_1}{V_2}\right]^\gamma = 1\left[\frac{V}{V/2}\right]^{1.4} = 1\,(2)^{1.4}$$

$$= 2.64 \text{ Atmosphere.}$$

6. ***A piston cylinder assembly contains 32 gm. of oxygen at a pressure of 4 atmosphere and a temperature of 27°C. It undergoes a cycle in which it is first heated at constant volume until the pressure becomes three times, then it expand isothermally until the pressure drops to the original value and finally it is cooled isobarically back to the original temperature. Calculate the change in internal energy and the work done in each process. Given atomic weight of oxygen = 16 and R = 8.3 Joule/mole/K.***

Solution: (i) When gas is heated at constant volume from 4.0 atoms to 12.00 atoms, the process is isochoric and since volume remains constant, the work done in the process.

$$dW = P.dV = 0.$$

Hence, from 1st law of thermodynamics

$$dQ = dU + P.dV$$
$$= C_V\, dT + 0.$$

Now from $\frac{P_1V_1}{T_1} = \frac{P_2V_2}{T_2}$

$$\frac{P}{300} = \frac{3P}{T_2} \qquad (\because V_1 = V_2)$$

$\therefore$ $T_2 = 900K$

Hence $dT = 900 - 300 = 600$ K.

Also for a diatomic gas like oxygen.

$$C_p = \frac{7}{2}R = \frac{7}{2}\times\frac{8.3}{4.18} = 7\,\text{cal/mol.K}$$

Hence $$C_V = \frac{C_p}{y} = \frac{C_p}{1.4} = 5\,\text{cal/mol.K}$$

$\therefore$ $dQ = dU = C_V\, dT = 5 \times 500 = 3000$ cals.

(ii) When the gas expands isothermally from 12 atmosphere to 4 atoms, the temperature remains constant and hence

$$dU = 0$$

Also for an isothermal process

$$dW = 2.3026\, RT \log \frac{V_2}{V_1}$$

Since $$P_1 V_1 = P_2 V_2$$

$$\frac{V_2}{V_1} = \frac{P_1}{P_2} = \frac{12}{4} = 3$$

$\therefore$ $$dW = 2.3026 \times \frac{8.3}{4.18} \times 900 \log_{10} 3$$
$$= 1977 \text{ cals.}$$

(ii) When the gas is cooled isobarically (constant pressure)

$$dU = c_V\, dT = 5\ (300 - 900)$$

$$= -\ 3000 \text{ cals.}$$

Also $dQ = Cp\ dT = 7\ (300 - 900) = -\ 4200$ cals.

$$\therefore \quad dW = dQ - dU$$

$$= -\ 4200 + 3000 = -\ 1200 \text{ cals.}$$

$$= -\ 1200 \text{ cals.}$$

7. A monoatomic ideal gas of volume 1 litre and pressure 8 atmosphere undergoes adiabatic expansion until the pressure drops to 1 atmosphere. What is the final volume and how much work is done ? Given 1 atoms = 10^5 N/m^2.

Solution: During an adiabatic process

$$P_1 V_1^{\gamma} = P_2 V_2^{\gamma}$$

Here, $P_1 = 8$ atoms. $= 8 \times 10^5$ N/m^2

$V_1 = 1$ liter $= 1000$ c.c. $= 10^{-3}$ m^3

$P_2 = 1$ atoms $= 10^5$ N/m^2

$V_2 = ?$

and for a monoatomic gas, the specific heat ratio y = 1.66.

$$\therefore \quad V_2 = \left(\frac{P_1}{P_2}\right)^{1/\gamma} V$$

$$= \left(\frac{8\times10^5}{1\times10^5}\right)^{1/1.66} \times 10^3$$

$$= (8)^{1/1.66} \times 10^{-3}$$

or $\quad V_2 = 3.47 \times 10^{-3}$ m^3

The work done during the adiabatic process from eq. (ii) § 1.21

$$W = \frac{1}{1-\gamma}[P_1V_2 - P_1V_1]$$

$$= \frac{1}{1-1.66}[1\times10^5 \times 3.47\times10^{-3} - 8\times10^5 \times 10^{-3}]$$

$$= \frac{1}{0.66}\ [4.53 \times 10^2] = 686 \text{ Joules.}$$

8. *A quantity of air (y = 1.4) at 27°C is compressed (a) slowly and (b) suddenly to one third of its volume. Find the change in temperature in each case.*

Solution: *(a) When the air is compressed slowly, the heat generated conducts away to the surroundings. Thus it is an isothermal process, temperature remains constant throughout, i.e., there is no change in temperature.*

(b) When the air is compressed suddenly, the process is adiabatic.

Hence $\quad T_1 V_1^{\gamma} -$ = constant.

If T_1, V_1 be the initial and T_2, V_2 the final temperature and volume respectively, then

$$T_1 V_1^{\gamma-1} = T_2 V_2^{\gamma-1}$$

or
$$\frac{T_1}{T_2} = \left(\frac{V_2}{V_1}\right)^{\gamma-1}$$

Here, T_1 = 27°C = (27 + 273) K = 300 K, $\frac{V_2}{V_1} = \frac{1}{3}$ and Y = 1.4

Thus
$$\frac{300}{T_2} = \left(\frac{1}{3}\right)^{1.4-1}$$

or
$$T_2 = 300\ (3)^{0.4}$$

$$= 465.5 \text{ K} = (465.5 - 273) = 192.5°\text{C}$$

$\therefore$ Change in temperature $T_2 - T_1$ = 192.5°C − 27°C

$$= 165°\text{C}.$$

9. *A motor car tyre has a pressure of 2 atmosphere at a room temperature of 27°C. If the tyre suddenly bursts, find the resulting temperature. (y for air = 1.40).*

Solution: Here, Pressure before burst, P_1 = 2 Atmospheres

Temperature before burst, T_1 = 27°C = (273 + 27) K = 300 K

Pressure after burst, P_2 = 1 Atmosphere

Temperature after burst T_2 = ?

For a sudden (adiabatic) expansion

$$\frac{P_1^{\gamma-1}}{T_1^{\gamma}} = \frac{P_2^{\gamma-1}}{T_2^{\gamma}}$$

or $$\left(\frac{T_2}{T_1}\right)^{\gamma} = \left(\frac{T_2}{T_1}\right)^{\gamma}$$

or $$\left(\frac{T_2}{300}\right)^{1.4} = \left(\frac{1}{2}\right)^{1.4-1}$$

or $$T_2 = 246.1 \text{ K}$$

$$= (246.1 - 273)°\text{C} = -26.9°\text{C}.$$

10. *A fixed mass of air at 1 atmosphere pressure is compressed adiabatically to 5 atmosphere and allowed to expand isothermally to its original volume*

(i) Represent these changes on a P – V diagram.

(ii) Calculate the pressure at the end of the isothermal process.

Given γ for air = 1.4.

Solution:

(i) The changes, the substance undergoes in the process of compression and expansion, are represented by a *P – V* curve shown in Fig. Curve *AB* represents the adiabatic compression from 1 atoms. to 5 atoms, while curve *BC* represents the isothermal expansion to its original volume. It is to be noted that the adiabatic curve is steeper than the isothermal curve.

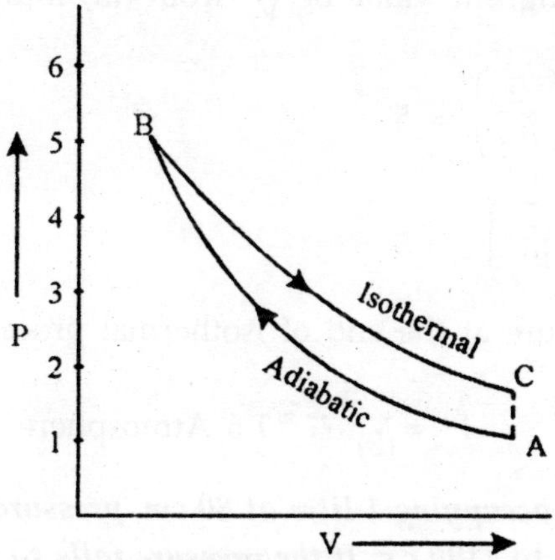

(ii) During an adiabatic process

$$P_1 V_1^{\gamma} = P_2 V_2^{\gamma}$$

Here $P_1 = 1$ atmosphere and $P_2 = 5$ atmosphere

$\therefore$ $1\ V_1^{\gamma} = 5\ V_2^{\gamma}$

or $\left(\dfrac{V_1}{V_2}\right)^{\gamma} = 5$ ... (i)

Now during an isothermal process

$$P'_1\ V'_1 = P'_2\ V'_2 \qquad \text{... (ii)}$$

As the gas expands from a pressure of 5 atmospheres and volume V_2 to its original volume V_1

$$P'_1 = 5 \text{ Atoms.}, V'_1 = V_2$$

$$V_2' = V_1, P'_2 = ?$$

Therefore, equation (ii) becomes

$$5 \,.\, V_2 = P'_2 \,.\, V_1$$

or $$\frac{V_1}{V_2} = \frac{5}{P'_2} \qquad \text{... (iii)}$$

Substituting, the value of $\frac{V_1}{V_2}$ from (iii) into (i), we get

or $$\left(\frac{5}{P'_2}\right)^{\gamma} = 5$$

$$\left(\frac{5}{P'_2}\right)^{1.4} = 5$$

or Pressure at the end of isothermal process

$$P'_2 = \frac{5}{(5)^{1/1.4}} \approx 1.6 \text{ Atmosphere.}$$

11. ***A gas occupying 1 litre at 80 cm. pressure is expanded adiabatically to 1190 c.c. If the pressure falls to 60 cm. in the process, deduce the value of*** γ.

Solution: Here V_1 = 1 liter = 1000 cm³ = 10^{-3} m³,

P_1 = 80 cm. = 0.8 *m*.

V_2 = 1190 × *10-6 m³*

P_2 = 0.6 m.

For an adiabatic change

$$P_1 V_1^{\gamma} = P_2 V_2^{\gamma}$$

or $$\left(\frac{V_1}{V_2}\right)^{\gamma} = \frac{P_2}{P_1}$$

Hence $$\left(\frac{10^{-3}}{1190 \times 10^{-6}}\right)^{\gamma} = \frac{0.6}{0.8}$$

or $$\gamma = \frac{\log 0.6 - \log 0.8}{\log 1000 - \log 1190}$$

$$= \frac{0.7782 - 0.9031}{3.0000 - 3.0755} = \frac{0.1249}{0.0755} = 1.65\ .$$

12. A *monoatomic ideal gas initially at 27°C is suddenly compressed to one tenth of its original volume. Calculate its temperature after compression. Make the same calculations for a diatomic gas like oxygen.*

Solution: When the gas is compressed suddenly, the process is adiabatic. Hence,

$$T_1 V_1^{\gamma-1} = T_2 V_2^{\gamma-1}$$

or $$T_2 = T_1\left(\frac{V_1}{V_2}\right)^{\gamma-1}$$

where T_1, T_2 and V_1, V_2 are respectively the initial and final temperatures and volume.

Here $T_1 = 27°C = (27 + 273)\ K = 300\ K$, $\frac{V_1}{V_2} = 10$,

$T_2 = ?$ and for a monoatomic gas

$\gamma = 1.66$.

∴ Temperature after compression

$$T_2 = 300\,(10)^{1.66-1} = 300 \times (10)^{0.66}$$

or $$\log T_2 = \log 300 + 0.66 \log 10$$

$$= 2.4771 + 0.66 \times 1.00 = 3.1371$$

∴ $$T_2 = 1371\ K = 1371 - 273 = 1098°C$$

and for a diatomic gas like oxygen, $\gamma = 1.4$

∴ $$T_2 = 300\,(10)^{1.4-1} = 300\,(10)^{0.4}$$

or $$\log T_2 = \log 300 + 0.4 \log 10$$

$$= 2.4771 + 0.4 \times 1 = 2.8771$$

∴ Temperature after compression,

$$T_2 = 753.6K$$

$$= 753.6 - 273 = 480.6°C.$$

Cooling Due to Adiabatic Reversible Processes : On the basis of first law of thermodynamics, it is possible to show that when a thermodynamic system such as a gas, stretched wire or s paramagnetic salt is made to undergo an adiabatic process reversibly the temperature of the system changes. The complete thermodynamic coordinates of these systems are:

1. *Gas Expansion:* The pressure of the gas (P), the volume of the gas (V) and the temperature of the gas (T) *i.e.*, this system is a function of the three parameters (P, V, T).

2. *Stretched Wire :* The stretching force (F), the length of the stretching wire (l), and the temperature of the wire (T), *i.e.*, the system is a function of parameters (F, l, T). The pressure and volume are considered here constant.

3. *Paramagnetic Salt:* The magnetic field (H), the intensity of magnetisation (I) and the temperature of the salt (T) *i.e.*, the thermodynamic coordinates are (H, I, T).

In all these examples, we can calculate the work done (dW) in a reversible infinitesimal change from these parameters of state's. For the various systems that we are considering, the work done may be shown to be:

1. *Gas Expansion :* Work done by the system is
   $dW = P.dV$

2. *Stretching of Wire :*
   $dW = -F\,dl$

3. *Paramagnetic Salt in a Magnetic Field :*
   $dW = -H\,dI.$

These give the work done by the system is each case. The convention adopted here is that the work done by the system is taken positive. For example, in case of gas expansion, the gas has to do work against a pressure that is acting on it while in case of stretching of wire, the work has been done by tension

F for a change of length dl because the wire by itself wants to contract and not expand as opposed to gas. Hence, a negative sign is used with $F\,dl$ and similar is the case with the last one.

A study of these expressions for dW shows that the work done in all these examples may be conveniently represented by a generalised formula,

$$dW = Xdx$$

where X may be called as the generalised force and x the generalised extension. The generalised force $(P, -F, -H)$ is a unique function of the extensions (V, l, I) respectively at a fixed temperature. These generalised parameters for these process assume the values:

1. Gas expansion : $X = P$, $x = V$
2. Stretching of wire : $X = -F$, $x = l$
3. Paramagnetic salt in field : $X = -H$, $x = I$.

By studying any system in detail, it is possible to derive a generalised result and then to apply this result to obtain the results of other systems directly by a proper substitution of generalised parameters. Let us try to achieve this.

We have from first law of thermodynamics

$$dQ = dU + dW.$$

In case of an adiabatic process, no heat is supplied or removed

i.e., $\quad dQ = 0.$

Hence $\quad dU = -dW.$

In terms of generalised parameters, it may be written as

$$dU = -Xdx.$$

If during the adiabatic process, the temperature of the system changes by dT and the thermal capacity of the system at constant volume is C_V, then

$$dT = -\frac{1}{C_V} X\,dx \quad (\because\ dU = C_V dT) \qquad \text{...(ii)}$$

Let us now consider the process of adiabatic expansion of a gas in some detail and find out an expression for the change in temperature. As in this case $dW = X\,dx = P\,dV$ and hence the change in temperature

$$dT = -\frac{1}{C_V} P.dV. \qquad \text{...(iii)}$$

But during an adiabatic process, the pressure and volume are governed by Poisson's law

$$PV^{\gamma} = \text{constant}.$$

Differentiating it,

$$\gamma\, PV^{\gamma-1}\, dV + V^{\gamma} P = 0$$

or
$$P\ dV = -\frac{V}{\gamma} dP$$

But for n gm. moles of a gas, the equation of state is

$$PV = nRT$$

which on constant pressure provides

$$P\left(\frac{\partial V}{\partial T}\right)_p = nR$$

or
$$PT\left(\frac{\partial V}{\partial T}\right)_p = nRT = PV$$

or
$$T\left(\frac{\partial V}{\partial T}\right)_p = V$$

Substituting this value of V in equation (iv)

$$P.dV = -\frac{T}{y}\left(\frac{\partial V}{\partial T}\right)_p dP.$$

Hence, the change in temperature of the gas is given by equation (iii) using this value of $P\ dV$, to be

$$dT = \frac{T}{\gamma C_V}\left(\frac{\partial V}{\partial T}\right)_p dP$$

$$= \frac{T}{C_p}\left(\frac{\partial V}{\partial T}\right)_p dP \qquad \left(\because \frac{C_p}{C_V} = \gamma\right). \quad \ldots(v)$$

In terms of generalised coordinates, the change in temperature during adiabatic process may, therefore, be written as

$$dT = \frac{T}{C_X}\left(\frac{\partial x}{\partial T}\right)_x dX. \quad \ldots(vi)$$

Knowing this equation of state of the system in terms of generalised parameters, the change in temperature for other systems can be obtained directly from this relation.

1. *Adiabatic expansion of gas.*

In case of a gas,

$$X = P \text{ and } x = V.$$

Hence, change in temperature is given from (vi) to be

$$dT = \frac{T}{C_p}\left(\frac{\partial V}{\partial T}\right)_p dP. \quad \ldots(vii)$$

Since in the expansion of the gas, the pressure of gas after expansion is lower as compared to the pressure before expansion *i.e.*, during expansion the pressure of the gas falls, and hence *the temperature of the gas also decreases*. Thus, if highly compressed gas be suddenly allowed to expand adiabatically, it cools very considerably. It is on account of the work done by the gas during expansion for which it draws upon its own reserve of internal energy (since no heat supply comes from outside) with a consequent fall of temperature.

If now, a be the coefficient of volume expansion of the gas, then

$$a = \frac{1}{V}\left(\frac{\partial V}{\partial T}\right)_p$$

and hence the change in temperature during adiabatic expansion of the gas is given by

$$dT = \frac{TV\alpha}{C_p} dP. \qquad \text{...(viii)}$$

2. *Stretching of a Wire*

In case of adiabatic stretching of a wire

$$X = -F \text{ and } x = l.$$

Substituting these values in (vi), we get the change in temperature

$$dT = \frac{T}{C_p}\left(\frac{\partial l}{\partial T}\right)_F dF \qquad \text{... (ix)}$$

Since the length of wire increases with temperature, $\left(\frac{\partial l}{\partial T}\right)_F$ is positive and therefore it is obvious from this expression that the wires *of those substances which expand on heating will show a cooling effect when stretched adiabatically.*

If β is the coefficient of linear expansion, m the mass per unit length of the wire and Cp the specific heat in mechanical units, then

$$\beta = \frac{1}{l}\left(\frac{\partial l}{\partial T}\right)_F \text{ or } \left(\frac{\partial l}{\partial T}\right)_F = \beta l$$

and $\qquad C_F = m\, l\, C_P$

therefore $\qquad dT = -\frac{T}{mlC_p}\beta l \; dF = -\frac{T\beta}{mC_p} dF \qquad \text{...(x)}$

This expression gives the fall in temperature due to adiabatic stretching of wire.

3. *Adiabatic Demagnetisation of Paramagnetic Salt*

In this case,

$$X = -H \text{ and } x = I.$$

Hence from (vi) the change in temperature is given by

$$dT = -\frac{T}{C_H}\left(\frac{\partial l}{\partial T}\right)_H dH \qquad \ldots \text{(xi)}$$

where C_H is the specific heat of the substance at constant field.

It is found that $\left(\frac{\partial l}{\partial T}\right)_H$ is always negative so that the right hand side is positive. If *H* decreases, then *dH* is negative and so also *dT*. Hence, cooling is produced during adiabatic demagnetisation of a paramagnetic substance.

PROBLEMS

1. ***The pressure on 20 gm. of water at 0°C is increased reversibly and adiabatically from 0 to 500 atmosphere. Calculate the change in temperature if the coefficient of volume expansion of water = – 0.67 × 10^{-6} °C^{-1}, Cp = 1.0087 kilo cal. /kg °C and density of water = 1000 kg/m^3.***

Solution: From eq. (viii) of § 1.26, the change in temperature

$$dT = \frac{TV\alpha}{C_p} dP.$$

Here T = 0°C = 273 K, Volume $V = \frac{\text{mass}}{\text{density}} = \frac{20 \times 10^{-3}}{1000}$

$= 20 \times 10^{-6}\ m^3$

$\alpha = -0.67 \times 10^{-6}$ °C^{-1}, $Cp = 1.0087$ kilo cals Kg^{-1} °C^{-1}

$= 1.0087 \times 4.2 \times 10^3$ Joule $kg.^{-1}$ °C^{-1}

and $dP = 500$ atmosphere

$= 500 \times 1.01 \times 10^5\ N/m^2$.

Therefore change in temperature

$$dT = \frac{TV\alpha}{C_p} dP.$$

$$= -\frac{273 \times 20 \times 10^{-6} \times 0.67 \times 10^{-6} \times 500 \times 1.013 \times 10^{5}}{1.0087 \times 4.2 \times 10^{3}}$$

$$= -0.044 \text{ K} = -0.044 \text{ °C}.$$

2. *In an adiabatic expansion of one mole of hydrogen gas the initial and final pressures are 20 and 2 atmosphere respectively while the initial and final volumes are 1 ànd 5 liters. Calculate the fall in temperature of the gas if its initial temperature = 27°C; C_p = 6.87 cal. I mol; and coefficient of volume expansion a = 0.0036013 deg^{-1}.*

Solution : Given $dP = P_i - Pf = 20 - 2 = 18$ atmospheres

$= 18 \times 1.031 \times 10^5$ N/m^2

$V_1 = 1$ liter $= 10^{-3}$ m³, T = 27°C = 300 K

$\alpha = 0.0036013 \text{ deg}^{-1}$

$C_p = 6.87$ Calorie mole $^{-1}$.

$= 6.87 \times 4.2$ Joule mole^{-1}.

Hence, fall in temperature

$$dT = \frac{TV\alpha}{C_p} dP.$$

$$= -\frac{300 \times 10^{-3} \times 0.0036013 \times 18 \times 1.013 \times 10^{5}}{6.87 \times 4.2}$$

$$= 68.3\text{K} = 68.3\text{°C}.$$

3. *A steel bar 0.40 cm ^{2} in cross section and initially at 27°C is suddenly stretched by applying a stretching force 2 × 10^3 newtons. Calculate the temperature immediately after stretching. (For steel, density = 7.8 gm. cm^{-3}, specific heat = 0.42 Joule gm $^{-1}$ deg ^{1}, coefficient* of *linear expansion = 1.0 × 10^{-3} deg 1).*

Solution: The change in temperature of the wire is given by eq. (x) § 1.26 to be

$$dT = \frac{T\beta}{m\,C_p} dF$$

Here $T = 27 + 273 = 300$ K, $\beta = 1 \times 10^{-5}$ deg.$^{-1}$,

$Cp = 0.42 \times 10^3$ Joule kg.$^{-1}$ deg.$^{-1}$

Mass per unit length

$$m = 0.40 \times 10^{-4} \text{ m}^2 \times 7.8 \times 10^3 \text{ kg/m}^3$$

$$= 0.40 \times 7.8 \times 10^{-1} \text{ kg/m}$$

$$dF = 2 \times 10^3 \text{ newton}$$

$$dT = \frac{300 \times 1.0 \times 10^{-5} \times 2 \times 10^3}{0.4 \times 7.8 \times 10^{-1} \times 0.42 \times 10^3}$$

$$= -0.0458 \text{ K} = -0.0458°\text{C}$$

Hence, the temperature immediately after stretching

$$= 27 - 0.0458$$

$$= 26.9542°\text{C}.$$

4. ***The tension of a copper wire of mass of 3 gm., diameter 1 m.m. and temperature 0°C is increased adiabatically from 0 to 1000 newtons. Calculate the change in temperature of the wire. (For copper, density = 8.93 × 10³ kg/m³, coefficient of linear expansion = 16.8 × 10⁻⁶ deg⁻¹, specific heat = 0.091 cals. gm.⁻¹ deg.⁻¹).***

Solution: The change in temperature of the wire

$$dT = -\frac{T\beta}{mC_p} dF$$

Given $T = 273$ K, $\beta = 16.8 \times 10^{-6}$ deg.$^{-1}$

$$m = \pi \times (0.5 \times 10^{-3})^2 \times 8.93 \times 10^3$$

$$= \pi \times 0.25 \times 8.93 \times 10^{-3}$$

$$C_p = 0.091 \text{ cal. gm.}^{-1} \text{ deg.}^{-1}$$

$$= 0.091 \times 4.2 \times 10^3 \text{ Joule kg.}^{-1} \text{ deg.}^{-1}$$

$$dF = 1000 \text{ N}$$

$$\therefore \quad dT = \frac{273\times 16.8\times 10^{-6}\times 1000}{\pi\times 0.25\times 8.93\times 10^{-3}\times 0.091\times 4.2\times 10^{3}}$$

$$= -1.71\text{K}$$

$$= -1.71°\text{C}.$$

Joule's Law for a Perfect Gas

If a fixed mass of gas is allowed to expand without doing external work under the condition that no heat enters or leaves the gas i.e., the gas is allowed to undergo what is called *free expansion* then if the molecules of the gas attract one another, internal work will have to be done by the gas in pulling them apart and this will happen at the expense of internal kinetic energy (heat energy) of the gas, which will, therefore, be cooled by the expansion. On the other hand if the molecules repel one another, the expansion should be accompanied by a rise of temperature and if there are no intermolecular attractions, there should be no change in temperature.

Joule's Experiment

In order to test these predictions, Joule took two copper cylinders *A* and *B* joined by a tube fitted with a stop cock *S* (Fig. below). *A* contained dry air at high pressure while *B* was perfectly evacuated. Both cylinders were immersed in the minimum amount of water whose temperature was noted with a sensitive thermometer. The stop cock was opened, the air from cylinder *A*

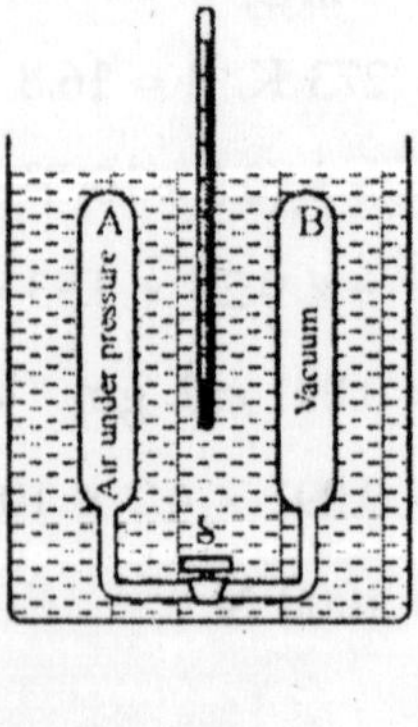

expanded into the vacuum in *B*, thus doing no external work (Free expansion).

The expansion was so rapid that no heat communication was possible with the surroundings and hence a cooling was expected if inter-molecular attractions existed. But Joule could not detect any change in the temperature of the water. He then varied his experiment by immersing *A* in *B* in separate water baths, but still no trace of intermolecular attractions was observed : the fall of temperature in *A* was exactly equal to the rise of temperature in *B*. If, however, inter-molecular attractions existed, the fall of temperature in *A* would have been greater than rise of temperature in *B*, the net loss accounting for the internal work done in pulling the molecules further apart.

Hence, the net conclusion was that no inter-molecular attractions existed in a gas or in other words, whatever changes may take place in the pressure and volume of a gas, so long as its temperature remains constant its internal energy remains unchanged. This result is often referred to as *Joule's law* and may be stated that *the internal energy (U) of a gas depends only on its temperature (T) and is quite independent of its volume (V) and pressure P*. Mathematically, it may be expressed by saying that *U* is a function of T only

i.e., $U = f\,(T).$

The following results follow directly from it

$$\left(\frac{\partial U}{\partial V}\right)_T = 0 \quad \text{and} \quad \left(\frac{\partial U}{\partial P}\right)_T = 0$$

and $$\left(\frac{\partial U}{\partial T}\right)_P = \left(\frac{\partial U}{\partial T}\right)_V .$$

The first equation tells that the internal energy of a system does not change with volume or pressure provided the temperature remains constant. The second equation states that the increase of internal energy per degree of temperature is the same no matter whether the pressure or the volume is kept constant in the process.

The Joule's law is obeyed strictly by a perfect gas only.

Departure from Joule's Law in a Vander-waal's Gas. ***(Vander Waal's Equation)*** : The Joule's law $\left(\frac{\partial U}{\partial V}\right)_T = 0$ is true for a perfect gas only as it is based on the assumption that no inter-molecular attractions existed in a gas. But the evidence in favour of the existence of inter-molecular attractions from other sides is quite strong. The equation of state $PV = RT$ deduced for an ideal gas was also based on the assumption that molecules are infinitestimal in size and collisions are perfectly elastic, there being no inter-molecular forces. But this equation has been inadequate to describe the behaviour of real gases as none of them is perfect. Experiments over a wide range of temperature and pressure showed that actual gases deviated from the perfect gas equation and the *deviation was quite large at high pressures and low temperatures*. It indicates that these deviations are due to mutual attractions between the molecules and the finite size of the molecules.

Joule Thomson experiment also established the existence of molecular attractions in gases. Actually at high pressures the size of the molecules of the gas becomes significant and can not be neglected. Also at high pressures, the molecules come nearer and the forces of inter-molecular attractions are appreciable. Vander Waal, therefore, corrected the perfect gas equation allowing for molecular attractions and the finite volume occupied by the molecules.

Correction for the Inter-molecular Attractions

A molecule in the interior of gas is attracted by other molecules in all directions and hence the net cohesive force on it is zero. But when a molecule strikes the walls of the container, it is pulled back by other molecules and hence the observed pressure of the gas at the walls of the container is less than the actual pressure within the body of the gas. This decrease in pressure P is proportional to (i) the number of attracting molecules per unit volume, and (ii) the number of attracted molecules striking per unit area of the wall of the container per second. Both these factors depend upon the number of molecules per c.c. or the density p of the gas.

$$\therefore \quad \text{Decrease in pressure } p \propto \rho^2 \propto \frac{1}{V^2}\eta$$

$$\text{or} \qquad p = \frac{a}{V^2}$$

where a *is* a constant and V is the volume of the gas.

$$\text{Hence, real pressure } = (P + p) = \left(P + \frac{a}{V^2}\right)$$

where P is the observed pressure.

Therefore, instead of P, we shall write in the equation of state the expression $\left(P + \frac{a}{V^2}\right)$

Correction for the Finite Size of the Molecules

As the molecules have finite size they occupy a space. Hence, the actual space for their movement is less than the observed volume V of the container. Therefore, the volume V in the equation of state be written as $(V - b)$ where b is a constant for unit mass of the gas.

Thus, the Vander Waal's equation of state for a gas is

$$\left(P + \frac{a}{V^2}\right)(V - b) = RT$$

where a and b are known as Vander Waal's constant and are different for different gases.

Discussion : It should be appreciated that Vander Waal's equation of state includes qualitatively all features of actual gases. It is a convenient way to describe the behaviour of real gases and succeeds in explaining the deviations from the perfect gas equation. But *over very wide pressure and temperature ranges it fails*. It breaks down when applied to the differences between the various gases. The values of parameters *a* and *b* do not remain constant for a particular gas but vary considerably at different temperatures. Also their values obtained by different methods do not agree. Moreover, if Vander Waal's equation is correct, the critical coefficient $\left(\frac{RT_C}{P_C V_C}\right)$ should have the same value 8/3 *(refer to chapter IV)* for all gases. Experimentally, however, it is found to vary form gas to gas and is greater than the theoretical value for all the gases, ranging from 3.13 for helium to 4.30 for water. The value of critical volume V_C is also found to vary experimentally from *1.4b* for argon to *2.3b* for hydrogen while according to theory $V_C = 3b$ for all the gases. Thus, Vander Waal's equation is not obeyed exactly by any gas in the neighbourhood of its critical point. However, it is still a very convenient equation over a wide range.

3

The Expression

Joule's Coefficient

Let us now discuss the *free expansion of a real gas* in Joule's experiment. In a real gas there exist inter-molecular forces and hence when the gas is allowed to expand, the inter-molecular separation changes and the gas has to do some internal work against the inter-molecular forces in separating the molecules further away from one another. This increases the potential energy of the gas and since no heat is allowed to enter or leave the gas (due to rapid expansion), it will happen at the cost of its kinetic energy.

Thus, the kinetic energy of the gas decreases resulting in a fall of its temperature. In this way, if intermolecular forces are attractive, a real gas will show a cooling effect when allowed to suffer a free expansion. On the similar reasoning, we may say that if inter-molecular forces are repulsive, the expansion will result in the heating of the gas. But, experimentally, it is seen that Joule's expansion of a real gas always produces cooling and hence the inter-molecular forces in a gas are attractive in nature. This change in temperature during free expansion of the gas is defined as *Joule's effect.*

Let 1 kg. mole of a gas undergo free expansion.

If δT is the change in temp. corresponding to change in volume δV, then Joule's coefficient is defined as

$$\eta = \left(\frac{\partial T}{\partial V}\right)_u$$

Now from thermodynamical treatment it can be shown that the variation of internal energy U of a gas with its volume V at any temperature T is given by

$$\left(\frac{\partial U}{\partial V}\right)_T = T\left(\frac{\partial P}{\partial T}\right)_V - P. \quad \text{... (ii)}$$

But according to Vander Waal's equation of state for real gas (or Vander Waal's gas).

$$\left(P + \frac{a}{V^2}\right)(V - b) = RT$$

or
$$P = \frac{RT}{V-b} - \frac{a}{V^2} \quad \text{...(iii)}$$

and hence
$$\left(\frac{\partial U}{\partial T}\right)_V = \frac{RT}{V-b} \quad \text{...(iv)}$$

Substituting the values of P and $\left(\frac{\partial U}{\partial T}\right)_V$ in eq. (ii), we get

$$T\left(\frac{\partial U}{\partial T}\right)_V - P = T.\frac{T}{V-b} - \frac{RT}{V-b} + \frac{a}{V^2} = \frac{a}{V^2} \quad \text{... (v)}$$

It indicates that the internal energy U is also a function of volume in addition to temperature. Hence, for a real gas

$$U = f\,(T, V) \quad \text{...(vi)}$$

Hence
$$dU = \left(\frac{\partial U}{\partial V}\right)_T dV + \left(\frac{\partial U}{\partial T}\right)_V dT$$

But during a free expansion, the internal energy of a gas remains constant, i.e., $dU = 0$

Hence $$\left(\frac{\partial U}{\partial V}\right)_T dV + \left(\frac{\partial U}{\partial T}\right)_V dT = 0$$

* From first law of thermodynamics

$$dQ = dU + dW = dU + PdV \qquad ...(1)$$

From second law of thermodynamics in terms of entropy

$$dQ = TdS \qquad ...(2)$$

Combining (1) and (2),

$$dU + PdV = TdS$$

or $$dU = TdS - PdV$$

This equation for isothermal changes gives

$$\left(\frac{\partial U}{\partial V}\right)_T = T\left(\frac{\partial S}{\partial V}\right)_T - P \qquad ...(3)$$

Maxwell's second relation is

$$\left(\frac{\partial S}{\partial V}\right)_T = \left(\frac{\partial P}{\partial V}\right)_V \qquad ...(4)$$

Using it, equation (3) gives

$$\left(\frac{\partial U}{\partial V}\right)_T = T\left(\frac{\partial P}{\partial V}\right)_{V-P} \qquad ...(5)$$

or $$\left(\frac{\partial T}{\partial V}\right)_u = -\frac{\left(\frac{\partial U}{\partial V}\right)_T}{\left(\frac{\partial U}{\partial V}\right)_V} \qquad ...(vii)$$

But $\left(\frac{\partial U}{\partial T}\right)_V$ is defined as the specific heat C_V of the gas at constant volume. Hence

$$\left(\frac{\partial T}{\partial V}\right)_U = -\frac{1}{C_V}\left(\frac{\partial U}{\partial V}\right)_T$$

The value of Joule's coefficient η is obtained by substituting in the above equation the value of $\left(\frac{\partial U}{\partial V}\right)_T$ from (v). Thus

$$\eta = \left(\frac{\partial T}{\partial V}\right)_U = -\frac{1}{C_V}.\frac{a}{V^2} \qquad \text{... (viii)}$$

Alternative Method : The change in temperature in free expansion of a gas is expressed in terms of Joule's coefficient defined by

$$\eta = \left(\frac{\partial T}{\partial V}\right)_U$$

where ∂T is the change in temperature due to change in volume ∂V in free expansion of the gas.

Now in Vander Waal's gas an internal pressure $P = \frac{a}{V^2}$ develops due to intermolecular attractions. Internal work done in Joule's expansion against this internal pressure

$$dW = p\ .\ dV = \frac{a}{V^2}dV$$

where α is Vander Waal's constant and V is initial volume of the gas. Consequently, internal kinetic energy decreases and internal potential energy increases in free expansion. Therefore,

Decrease in internal KE. = Increase in internal P.E.

$$= -dW = -\frac{a}{V^2}dV \qquad \text{... (i)}$$

Due to decrease in internal kinetic energy in free expansion, the temperature of the gas decreases. If this decrease in temperature is dT and C_v is the molar specific heat at constant volume, then

Decrease in internal K.E. $= C_V.dT$ ... (ii)

Equating (i) and (ii), we have

$$C_V.dT = -\frac{a}{V^2}dV$$

or $$\frac{dT}{dV} = -\frac{1}{C_V}.\frac{a}{V^2}$$

Therefore, Joule coefficient

$$\eta = \left(\frac{\partial T}{\partial V}\right)_U = -\frac{1}{C_V}.\frac{a}{V^2} \quad \text{... (iii)}$$

Discussion : It is clear from this equation that Joule's coefficient for a real gas is negative because 'a' *is* a positive quantity. Thus *when a real gas is allowed to suffer Joule's free expansion, a cooling effect is produced* which is in quite agreement with the experimental results. Equations (v) and (vii) represent the deviations of a Vander Waal's gas from Joule's law because they indicate that internal energy in case of a gas obeying Vander Waal's equation is a function of volume as opposed to a perfect gas whose internal energy is a function of temperature only. This departure of Vander Waal's gas from Joule's law, as we have seen, is due to the inter-molecular forces present within the gas. These forces are called Vander Waal's forces and are discussed in detail in the next article. Had these forces been absent, as in the case of an ideal or perfect gas, the gas will show no change in temperature on suffering a free expansion.

PROBLEMS

1. For a gas, the Vander Waal's constants are: a = 1.32 liter^2 – atoms, mole.$^{-2}$, b = 3.12 × 10^{-2} litre mole Calculate the temperature at which 5 mole gas at 5 atoms pressure will occupy 20.0 litre volume. Calculate also the pressure of this gas at the same temperature when volume is reduced to 2.0 litre. Given R = 8.4 Joules mole.$^{-1}$ deg.$^{-1}$.

Solution: The Vander Waal's equation of state is

$$\left(P+\frac{a}{V^2}\right)(V-b)=RT$$

For 5 molecules of gas, it becomes

$$\left(P+\frac{a}{V^2}\right)(V-b)=5RT$$

or
$$T=\frac{\left(P+\frac{a}{V^2}\right)(V-b)}{5R} \quad \text{...(i)}$$

Here α = 1.32 liter^2 atoms. mole^{-2}

= 1.32 × $(10^3)^2$ × 10^{-12} × 10^5 Nm^4 mole.$^{-2}$

= 0.132 Nm^4 mole.$^{-2}$

Because 1 liter = 10^3 cm.^{3} = 10^{-3} m^3

and 1 atmosphere ≈ 10^5 (N/m^2)

Also b = 3.12 × 10^{-2} liter mole^{-1}

= 3.12 × 10^{-2} × 10^{-3} = 3.12 × 10^{-5} m^3 mole.$^-$

Pressure, P = 5 atmosphere = 5 × 10^5 NM^{-2}

Volume V = 20.0 liter = 20 × 10^3 cm.^{3}

= 20 × 10^{-3} m^3

Gas constant $R = 8.4$ Joules mole.$^{-1}$ deg.$^{-1}$, T = ? substituting these values in equation (i)

$$T=\frac{\left\{5\times10^5+\frac{0.132}{\left(20\times10^{-3}\right)^2}\right\}\left\{20.0\times10^{-3}-3.12\times10^{-5}\right\}}{50\times8.4}$$

or
$$T=\frac{(500000+330)(0.01997)}{5\times8.4}$$

= 237.9K.5

Now when volume is reduced from 20.0 liter to 2.0 liter, *i.e.*, 1/10 of its original volume, the pressure will increase 10 times at the same temperature (because the product of P and V remains constant). Hence when volume is reduced to 2.0 liter.

The pressure of the gas = Initial pressure × 10

$$= 5 \times 10 = 50 \text{ atmosphere.}$$

Example: *Vander Waal's constant of a gas are a = 0.245 litre² atoms Mole⁻², b = 2.67 × 10⁻¹⁰ litre mole⁻¹. The initial volume and pressure of the gas are 2 litre and 4 atmosphere respectively. The gas is made to suffer a free expansion to triple its volume. Find the change in temperature of the gas. Given* C_v *= 5 cals. deg.⁻¹ mole⁻¹.*

Solution: The Joule's coefficient for a real gas is given by

$$\eta = \left(\frac{\partial T}{\partial V}\right)_u = -\frac{1}{C_V}\cdot\frac{a}{V^2}$$

or $$dT = -\frac{1}{C_V}\cdot\frac{a}{V^2}dV$$

Here $$\alpha = 0.245 \text{ liter}^2 \text{ atoms mole}^{-2}$$

$$= 0.245 \times (10^3)^2 \times 10^{-12} \times 10^5 \text{ Nm}^4 \text{ mole.}^{-2}$$

$$= 0.0245 \text{ Nm}^4 \text{ mole.}^{-2}$$

$$C_v = 5 \text{ cals. deg.}^{-1} \text{ mole.}^{-2}$$

$$V = 2 \text{ liter} = 2 \times 10^3 \text{ cm.}^3 = 2 \times 10^{-3} \text{ m}^3$$

$$dV = 2 \times 3 - 2 = 4 \text{ liter}$$

$$= 4 \times 10^3 \text{ cm.}^3 = 4 \times 10^{-3} \text{ m}^3$$

Hence change in temperature

$$dT = -\frac{1}{5\times 4.2}\times\frac{0.0245}{(2\times 10^{-3})^2}\times 4\times 10^{-3}$$

$$= -\frac{0.245\times 4\times 10^3}{5\times 4.2\times 4}$$

$$= -1.17 \text{ K} = -1.17°\text{C}.$$

Thus temperature will fall by 1.17°C.

Nature of Vander-waal's Forces

The matter is formed of atoms and molecules and has the tendency to occupy minimum space at low temperatures through transforming into liquid and solid states. From this, we may conclude that the atoms or molecules do not lose their individuality, inspite of this attractive force. We may further conclude that there must be repulsive forces at short intermolecular distances. The simplest model giving rise to such forces was assumed by Vander-Waal, according to which, "The molecules are hard and impenetrable surrounded by attractive field". This model does not explain the observed compressibility of solids and does not agree with the modern concept of electronic structure of atom suggested by Bohr and Rutherford according to which, "The atom consists of a positively charged nucleus surrounded by negatively charged electrons in shells". Hence, it may be assumed that the forces between atoms or molecules are due to interaction between the electrons and the nuclei of the two systems. Thus, we are led to the conclusion that intermolecular forces are essentially electrostatic. However, there are also magnetic forces which are ordinarily small and may be neglected. These forces of attraction are generally referred to *Vander Waal's forces.*

Except for atoms having closed shells such as Helium (He), Argon (Ar), Neon (Ne) etc., almost all other atoms are electrically unsymmetrical. Each of the such atoms are electrostatically equivalent to a positive and equal negative charge separated by a small distance i.e., an electric dipole possessing an electric moment equal to the product of the charge and distance between the two charges. The molecules formed of such atoms will also behave as dipoles. The dipole moment of an atom continuously varies in magnitude and direction. Therefore, the average dipole moment averaged over all phases of electronic motion of an atom is always zero. In the case of molecules, the average is not necessarily zero. If the average dipole moment of a molecule does not vanish, it will possess a permanent dipole moment.

Such molecules are called *polar molecules.* On the other hand if the average dipole moment of a molecule vanishes; the molecule is said to be *non-polar molecule.*

If the non-polar molecules are placed in an electric field, they have an average dipole moment and the molecule is said to be *polarised* while the dipole so formed is termed as *induced dipole.* On the other hand if the polar molecules are placed in an electric field, they possess changed dipole moment.

Thus, the Vander Waal's forces between the molecules of a matter (or gas), more conveniently the potential energy of the system, may be due to :

(i) Dipole-dipole interaction.

(ii) Dipole-induced dipole interaction.

(iii) Dispersion interaction.

These are discussed one by one below :

Dipole Dipole Interaction: Let there be two polar molecules having permanent diple moments p_1 and P_2. Let the effective charge of first dipole be $-q_1$ and q_1 separated by a distance $2l_1$, and those of the second dipole be $-q_2$ and q_2 separated by a distance $2l_2$.

Then we have

$$P = 2l_1 q_1$$

$$P_2 = 2l_2 q_2$$

$$AB = 2l_1, \; CD = 2l_2.$$

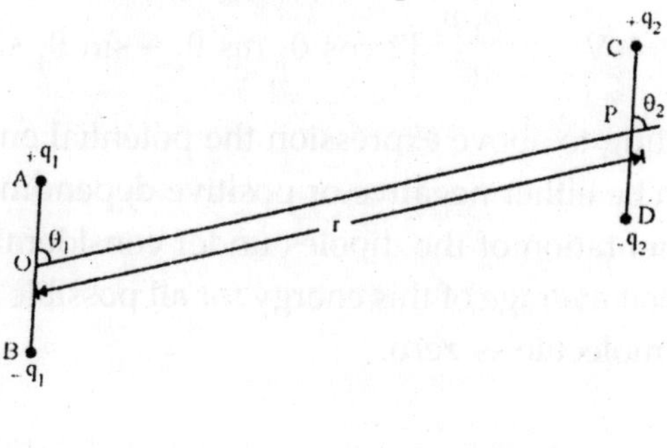

Let r be the internuclear distance, i.e., the distance between the centres of two dipoles and θ_1 and θ_2 the angles which the dipole makes with the internuclear axis as shown in Fig. above. Then taking O as origin, OP as x-axis and a line perpendicular to OP as y-axis, the Cartesian coordinates of points A, B, C and D are $(l_1 \cos \theta_1, l_1 \sin \theta_1)$, $(-l_1 \cos \theta_1, l_1 \sin \theta_1)$, $(r + l_2 \cos \theta_2, l_2 \sin \theta_2)$ and $(r - l_2 \cos \theta_2, - l_2 \sin \theta_2)$ respectively. So the distances between various charges, are given by

$$r_{AC} = [\{(r + l_2 \cos \theta_2) - l_1 \cos \theta_1)^2 + (l_2 \sin \theta_2 - l_1 \sin \theta_1)^2]^{1/2}$$

$$r_{AD} = [\{(r + l_2 \cos \theta_2) - l_1 \cos \theta_1)^2 + (l_2 \sin \theta_2 - l_1 \sin \theta_1)^2]^{1/2}$$

$$r_{BC} = [\{(r + l_2 \cos \theta_2) + l_1 \cos \theta_1)^2 + (l_2 \sin \theta_2 + l_1 \sin \theta_1)^2]^{1/2}$$

$$r_{BC} = [\{(r + l_2 \cos \theta_2) + l_1 \cos \theta_1)^2 + (-l_2 \sin \theta_2 + l_1 \sin \theta_1)^2]^{1/2}$$

The potential energy of the system is given by

$$V = \frac{q_1 q_2}{r_{AC}} - \frac{q_1 q_2}{r_{BC}} - \frac{q_1 q_2}{r_{AD}} + \frac{q_1 q_2}{r_{BD}}$$

$$= q_1 q_2 \left[\frac{1}{r_{AC}} - \frac{1}{r_{BC}} - \frac{1}{r_{AD}} + \frac{1}{r_{BD}} \right]$$

Substituting values of r_{AC}, r_{BC}, r_{AD} and r_{BD} in above equation and using Binomial expansion, we get

$$V = -\frac{4q_1 q_2 l_1 l_2}{r^3} [2 \cos \theta_1 \cos \theta_2 - \sin \theta_1 \sin \theta_2]$$

$$V = -\frac{p_2 p_1}{r^3} [2 \cos \theta_1 \cos \theta_2 - \sin \theta_1 \sin \theta_2]$$

According to above expression the potential energy of the system can be either negative or positive depending upon the relative orientation of the dipoles under consideration. It will be found that average of this energy for all possible orientation for either molecule is zero.

As $$F = -\frac{dV}{dr}.$$

Clearly the force varies as r^{-4}.

Dipole-Induced Diple Interaction : The field produced by an electric dipole of moment p at a distance r from its centre is given by

$$E = \frac{p}{r^3}\sqrt{(1+3\cos^2\theta)} \quad \text{...(1)}$$

where θ is the angle between $\vec{p}$ and $\vec{r}$.

When a molecule is placed in a field of strength E, the induced dipole moment p′ is given by

$$P' = aE$$

where α is the polarizability of the molecule.

The interaction energy of the system, i.e., the potential energy due to interaction between a permanent dipole of the moment p and induced dipole of moment p' is given by

$$V = -\int_0^E P' \, dE = -\int_0^E \alpha E \, dE$$

$$= \frac{1}{2}\alpha E^2 \cdot \quad \text{...(2)}$$

Here, we have taken into account the fact that the dipole moment p' is induced by the field E.

Substituting value of E from eqn. (1) in (2), we get

$$V = -\frac{1}{2}\alpha \cdot \frac{p^2}{r^6}(1+3\cos^2\theta)$$

As all the orientations of the dipole are possible, we must determine the average value of V which may be obtained directly if we substitute average value of

$$\cos^2\theta = \frac{1}{3}$$

Thus we have

$$V = -\frac{1}{2}\alpha \cdot \frac{p^2}{r^6}(1+1) = -\alpha \frac{p^2}{r^6}$$

Thus, the interaction energy between a dipole and induced dipole

varies as r^{-6}, and hence the force $\left(F = \frac{dV}{dr}\right) r^{-7}$ · varies as r^{7}.

Dispersion Interaction : In the case of a non-polar molecule, the average dipole moment averaged over all possible phases of electronic motion, is zero. But the instantaneous value of the dipole moment is never zero and has certain magnitude and direction. If another non-polar molecule is placed near the first, the instantaneous dipole moments of the two molecules will interact and hence will produce some potential energy which may be negative or positive. Such interaction between non-polar molecules is termed as *dispersion interaction*. The potential energy due to dispersion forces is given by

$$V = -\frac{3\,h v_0 \alpha^2}{4r^6}$$

where h is Planck's constant and V_0 is the frequency of the motion of electron.

From above discussion it is clear that the energy due to dipole interactions may be represented by $-Br^{-m}$, where B is a constant for the given substance and m in most cases is probably, equal to 6.

Exchange Forces : The above attractive forces operate when the atoms are sufficiently far apart; but when the atoms come so close together that they touch or overlap, quite new types of forces come into play. These forces are called *exchange forces*. For example, very low compressibilities, very dense gases and impenetrability of the atoms show that when molecules are brought sufficiently close together, the repulsive forces predominate and are of chief importance. The mutual repulsive potential energy as calculated from wave mechanics is equal to $De^{-\lambda r}$ for large distances r. For short distances the expression is very complicated and it has become the practice to utilize the simpler empirical formula $V = Ar^{-n}$, where A and n are constants. n *is* found to lie between 8 and 14.

Complete Potential Function

The complete potential function for the interaction of two molecules is the sum of Vander Waal's attractive potential and the exchange interaction and hence may be represented by

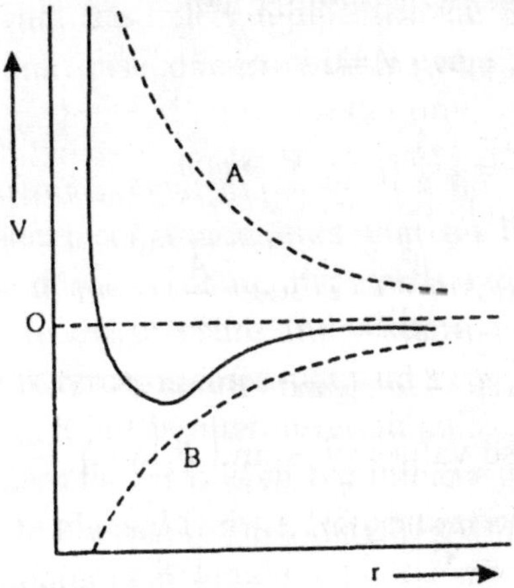

$$V = Ar^{-n} - Br^{m} \qquad \text{... (3)}$$

Here m is usually 6 and n is found to lie between 8 and 14. Hence, at moderate distance the attractive forces predominate; while at short ranges the repulsive forces predominate. The general nature of the potential function is represented in Fig. Curve A indicates the type of the plot for the repulsive term Ar^{-n}, while B shows the character of the attractive term $- Br^{-m}$. The resultant of the two terms is given by the full line curve.

Putting $n = 12$ and $m = 6$ in eq. (3),

$$V = Ar^{-12} - Br^{-6} \quad \text{... (4)}$$

$$\therefore \quad \frac{2\alpha}{Rb} = -12Ar^{-13} + 6Br^{-7}$$

When V→ minimum, $\frac{\partial V}{\partial r} = 0$ and $r = r_o$

$$\therefore \quad -12Ar_0^{-13} + 6Br_0^{-7} = 0$$

$$\therefore \quad r_0 = (2A/B)^{1/6} \quad \text{... (5)}$$

Substituting this value of r_0 in eq. (4), we get

$$V_{min} = -B^2/44$$

Now, it is evident from eq. (4),

At $\quad r = \infty, \; V = 0$

or At $\quad r = \left(\frac{A}{B}\right)^{1/6}, \quad V \to min$

If we suppose $-\frac{B^2}{44} = -\varepsilon$ and $\left(\frac{A}{B}\right)^{1/6} = \sigma$

On solving, $A = 4\varepsilon\sigma^{12}$ and $B = 4\varepsilon\sigma^{6}$

Putting these values of A and B in eq. (4), we get

$$V = 4\varepsilon\left[\left(\frac{\sigma}{r}\right)^{12} - \left(\frac{\sigma}{r}\right)^{6}\right]$$

This form of potential function is called Lenard Jones Potential.

PROBLEMS

1. ***The interaction energy between two molecules is given by $V_I = Ar^{-12} - Br^{-6}$ in a typical case where***

$A = 0.036 \times 10^{-104}$ erg. cm. ^{12} and $B = 11.7 \times 10^{-61}$ erg. cm.^{6}. Deduce (i) value of r at which $V_r = 0$, (ii) value of r at which V_r is minimum and (iii) the magnitude of $(V)_{r\,min}$ in electron volts.

Solution: (i) When $V_r = 0$, we have

$$Ar^{-12} - Br^{-6} = 0$$

or $$r = \left(\frac{B}{A}\right)^{-1/6} = \left(\frac{A}{B}\right)^{1/6}$$

Given A = 0.036 × 10^{-104} erg. cm.[12]

= 0.036 × 10^{-135} Joule m^{12}

and B = 11.7 × 10^{-61} erg. cm.[6] = 11.7 × 10^{-80} Joule m^6

$$\therefore \quad r = \left[\frac{0.036\times10^{-135}}{11.7\times10^{-80}}\right]^{1/6}$$

= 2.59 × 10^{-10} m = 2.59Å

(ii) For V_r to be minimum, $\frac{\partial Vr}{\partial r}$ must be equal to zero

or $$\frac{\partial Vr}{\partial r} = \frac{\partial}{\partial r}[Ar^{-12} - Br^{-6}] = 0$$

or $$-12Ar^{-13} + 6Br^{-7} = 0$$

or $$r = \left(\frac{2A}{B}\right)^{-1/6} = \left[\frac{2\times0.036\times10^{-135}}{11.7\times10^{-80}}\right]^{1/6}$$

= 2.916 x 10^{-10} M = 2.916Å

(iii) $$(V_r)_{min} = \frac{A}{r^{12}} - \frac{B}{r^6} = \frac{0.036\times10^{-135}}{(2.916\times10^{-10})^{12}} - \frac{11.7\times0^{-80}}{(2.916\times10^{-10})^6}$$

$$= -3.91 \times 10^{-23}\text{ Joule} = -\frac{3.91\times10^{-23}}{1.6\times10^{19}}eV$$

= – 2.44 x 10^{-4} eV.

Joule-thomson Expansion—The Porous Plug Experiment : Joule's experiment based on free adiabatic expansion of a gas failed to detect the existence of inter-molecular attractions. Joule however did not conclude that molecular attractions do not at all exist in a gas. He thought that his measurements were not sensitive enough to detect the small temperature change due to internal work that might have perhaps occurred during the experiment. Therefore, in 1952, in collaboration with Lord Kelvin (Thomson), he devised a very sensitive technique known as porous *plug experiment* which established fully the existence of inter-molecular attractions in the case of all actual gases.

The principle of Joule Thomson experiment consists in forcing a gas at a constant pressure through a porous plug thermally insulated from outside. The porous plug consists of a compressed porous material like cotton, wool or silk fibres having a number of fine holes or pores and is thus equivalent to a number of narrow orifices in parallel.

The gas, on passing through the plug under a large drop of pressure, gets throttled, i.e., its molecules get further apart from one another, so that the gas has to do some internal work against inter-molecular attractions, if these exist. This work will come from the internal energy of the gas since no heat is supplied to it. Therefore, the gas passing through the plug will suffer a fall in temperature.

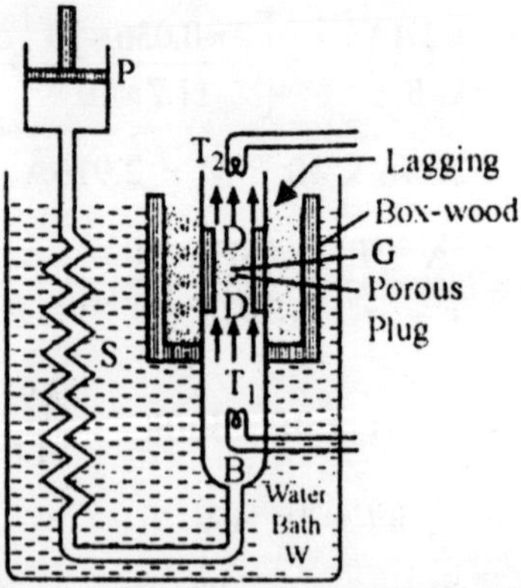

Since a cooling of the gas was actually demonstrated by Joule and Thomson in their famous porous plug experiment (indicating the existence of inter-molecular attractions) and hence it is called the *Joule-Thomson* (or Joule-Kelvin) *effect*. It is, however, entirely different from Joule-expansion where the gas is made to expand without doing any external work. Here in porous plug experiment, the gas expands against a constant external pressure so that it has to do external work or some work is done on it in addition to any internal work done against inter-molecular attractions.

The apparatus used in porous plug experiment is shown in Fig. The gas under test is compressed to a high pressure with the help of piston *P* and is passed through a long spiral tube *S* immersed in a water bath W maintained at a constant temperature. The compressed gas enters the tube *B* fitted with a porous plug *G* of cotton-wool or silk fibre. The plug is contained between two perforated brass discs *D, D*. A vessel *L* filled with cotton-wool surrounds the tube *B* to provide a good thermal insulation to the porous plug. When the gas flows through the plug, it suffers a large drop in pressure. When its flow becomes steady, the temperature of the gas on the two sides of the plug is measured with platinum resistance thermometers T_1 and T_2. The pressure of the incoming gas is measured by a manometer (not shown in Fig.) while the pressure of the out-going gas is equal to the atmospheric pressure.

The behaviour of a large number of gases was studied over different ranges of initial temperatures. It was found that in general all gases show a change in temperature in passing through the porous plug. This temperature change is referred to as *Joule Thomson effect*.

Theory of Joule-Thomson Expansion: Constancy of Enthalpy : Let us consider a simple arrangement shown in Fig. representing the porous plug experiment. CC′ is a cylinder with non-conducting walls having a porous plug *G* and fitted with two non-conducting pistons 1 and 2. A fixed mass of a gas having a volume V_1 is filled between the piston 1 and porous plug. Let this gas be highly compressed to a pressure P_1 and the pressure on other side of the plug be P_2, the piston 2 being kept just behind the plug preventing any gas from passing through it (Fig.). Due to this large difference of pressure, the gas flows through the porous plug and gets throttled or 'wire drawn' i.e., its molecules get further apart.

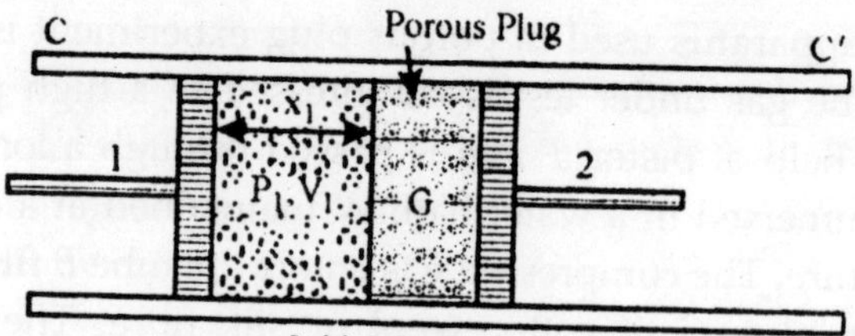

Initial State

Let now piston 1 be moved slowly inwards in such a way so as to keep a *constant pressure* P_1 on left hand side of the plug as the gas escapes. The piston 2 will move slowly outwards so as to keep a *lower constant pressure* P_2 on the right hand side of the plug. After all the gas has passed through the plug, the pistons 1 and 2 will be in the positions shown in Fig. The gas will now occupy a greater volume, say ${}_2$. If x_1 and x_2 are the distances through which pistons 1 and 2 move and A is the area of cross-section of the pistons, then

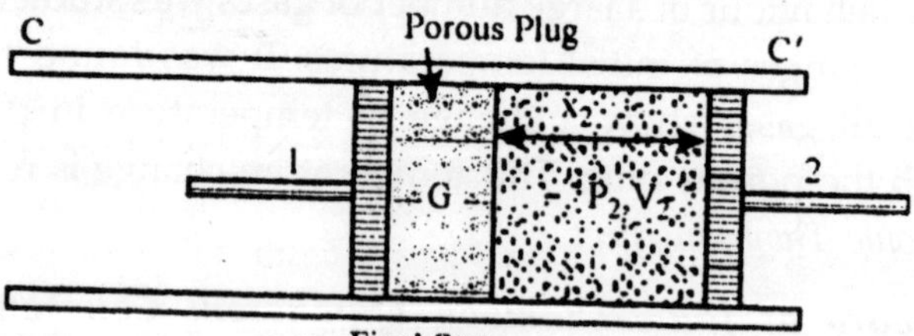

Final State

External work done on the gas by piston 1 $= P_1 \times A \times x_1$

$$= P_1 V_1$$

External work done by the escaping gas on piston 2

$$= P_2 \times A \times x_2$$
$$= P_2 V_2$$

$\therefore$ Net external work done by the gas in passing through the plug

$$= P_2 V_2 - P_1 V_1 \quad \text{...(i)}$$

As the cylinder is thermally insulated, no heat is exchanged between the gas and its surroundings. Therefore, the external work must have been done by the gas at the cost of its own internal energy. So that if U_1 and U_2 be the internal energies of the gas in the initial and final states respectively, the loss in its internal energy in passing through the plug is obviously, $U_1 - U_2$. Then according to first law of the thermodynamics, we have ($\therefore$ $dQ = 0$).

$$U_1 - U_2 = P_2 V_2 - P_1 V_1 \quad \text{... (ii)}$$

or $$U_1 + P_1 V_1 = U_2 + P_2 V_2 \quad \text{... (iii)}$$

or $$H_1 = H_2 \quad \text{... (iv)}$$

The quantity $(U + PV)$, which is conserved, is called enthalpy or total heat function and denoted by the letter H. Hence in a throttling process the initial and the final enthalpies are equal. One is, however, not entitled to say that the enthalpy remains constant because the gas escapes under the effect of a finite pressure and it is not possible to describe the state of the gas throughout the change. It is why the Joule-Thomson effect is sometimes referred to as the *iso-enthalpic effect.*

It is obvious from eq. (iii), if $P_2 V_2 > P_1 V_1$; $U_2 < U_1$ *i.e.*, the internal energy of the gas will decrease on passing through the porous plug and hence a cooling will result. If anyhow $P_2 V_2 < P_1 V_i$; $U_2 > U_1$ *i.e.* the internal energy will increase and a heating will result. If anyhow $P_2 V_2 = P_1 V_1$ (*i.e.* the gas is perfect and obeys Boyle's law). $U_1 = U_2$, the internal energy of the gas remains unchanged and so also its temperature.

PROBLEMS

1. ***Show that an isothermal curve for an ideal gas drawn on a P-V diagram is also isoenthalpic.***

Solution: As the internal energy (U) of a gas depends upon its temperature only (Joule's Law) : for an ideal gas along an isothermal curve, internal energy will remain constant, *i.e.*

$$U_i = U_f$$

Also for the ideal gas along isothermal curve

$$P_i V_i = P_f V_f \qquad \text{(Boyle's law)}$$

Thus $U_i + P_i V_i = U_f + P_f V_f$

or $\quad H_i = H_f$

i.e., the enthalpy H remains constant along an isothermal curve or in other words *the curve is also isoenthalpic.*

Vander Waal's Gas in Joule-thomson Expansion : As has already been mentioned, when a gas under constant pressure is forced through an insulated porous plug to a region of lower constant pressure, its temperature changes. This is called the Joule-Thomson effect. The magnitude and sign of this temperature change depends upon the initial temperature of the gas and is proportional to the difference of pressure between the two sides of the plug.

Let us suppose that one gm. mole of a real gas is allowed to expand through a porous plug from a pressure P_1 and volume V_1 to a pressure P_2 and volume V_2 (Refer to Figs.). During this process, suppose that the temperature also changes from T_1 to T_2 due to Joule-Thomson effect. The gas will have to do external as well as internal work. As it has been shown in the last article.

External work done on the gas $= P_1 V_1$

External work done by the gas $= P_2 V_2$

$\therefore$ Net external work done by the gas $= P_2 V_2 - P_1 V_1$

Now an internal work is also done by the gas in overcoming the forces of molecular attraction. In case of a Vander Waal's gas (gas obeying Vander Waal's equation), the attractive forces between the molecules of the gas are equivalent to an internal pressure $\frac{a}{V^2}$ where 'a' is a constant. Hence when 1 gm. mole of the gas expands from a volume V_1 to V_2, the internal work done by it is given by

$$\int_{V_1}^{V_2} P.dV = \int_{V_1}^{V_2} \frac{a}{V^2} dV = \frac{a}{V_1} - \frac{a}{V_2}.$$

Hence, the total work done by the gas

$$W = (P_2 V_2 - P_1 V_1) + \frac{a}{V_1} - \frac{a}{V_2} \quad \text{....(i)}$$

Now Vander Waal's equation of state for a gas is

$$\left(P + \frac{a}{V^2}\right)(V - b) = RT$$

or
$$PV + \frac{a}{V} - P.b - \frac{ab}{V^2} = RT$$

or
$$PV = RT + P.b - \frac{a}{V} \quad \left(\text{neglecting } \frac{ab}{V^2}\right) \quad \text{... (ii)}$$

Substituting the value of PV in equation (i), with appropriate suffixes, we get

$$W = \left(RT_2 + P_2 b - \frac{a}{V_2}\right) - \left(RT_1 + P_1 b - \frac{a}{V_1}\right) + \frac{a}{V_1} - \frac{a}{V_2}$$

$$= R(T_2 - T_1) - b(P_1 - P_2) + \frac{2a}{V_1} - \frac{2a}{V_2} \quad \text{... (iii)}$$

Since in equation (ii), *a* and *b* are very small, we may take approximately

$$PV = RT \quad \text{or} \quad V = \frac{RT}{P}$$

$$V_1 = \frac{RT_1}{P_1} \text{ and } V_2 \frac{RT_2}{P_2}$$

Substituting these values in (iii), we have

$$W = R(T_2 - T_1) - b(P_1 - P_2) + \frac{2\alpha P_1}{RT_1} - \frac{2\alpha P_2}{RT_2}$$

Now T_1 and T_2 are very nearly equal, we may write $T_1 - T_2 = dT$ and substitute T for T_1 or T_2. Hence

$$W = -Rdt - b(P_1 - P_2) + \frac{2\alpha}{RT}(P_1 - P_2)$$

$$= (P_1 - P_2)\left(\frac{2\alpha}{RT} - b\right) - Rdt \qquad \text{... (iv)}$$

As the gas is thermally insulated, the energy necessary for doing this work is drawn from the kinetic energy of the molecules. Hence, the kinetic energy decreases resulting in a fall of temperature by dT. If C_V is the specific heat of the gas at constant volume, then the heat lost by the gas to do the work W is $C_V\ dT$ *i.e.*, W is equivalent to $C_V\ dT$ cals. Hence

$$C_V dT = (P_1 - P_2)\left(\frac{2\alpha}{RT} - b\right) - R\,dt$$

$$\text{or} \quad dt(C_V + R) = (P_1 - P_2)\left(\frac{2\alpha}{RT} - b\right)$$

$$\text{Now} \quad C_P - C_V = RC_V + R = C_P$$

$$\therefore \quad C_P dt = (P_1 - P_2)\left(\frac{2\alpha}{RT} - b\right) \qquad \text{... (v)}$$

Hence the fall in temperature

$$dT = \frac{(P_1 - P_2)}{C_P} \cdot \left(\frac{2a}{RT} - b\right).$$

Thus, the cooling at any temperature is directly proportional to the fall in pressure $(P_1 - P_2)$ across the porous plug.

(i) Since $P_2 < P_1$ the quantity $(P_1 - P_2)$ is necessarily positive.

$$dT \text{ will be + ve if} \left(\frac{2a}{RT} - b\right) \text{ is +ve}$$

i.e. $$\frac{2\alpha}{RT} > b \quad \text{or} \quad T < \frac{2\alpha}{Rb}$$

Thus the cooling will take place if the initial temperature of the gas is less than $\frac{2\alpha}{Rb}$

(ii) dT will be zero if $\left(\frac{2\alpha}{RT} - b\right) = 0$

or $$T = \frac{2\alpha}{Rb}$$

This temperature is called the temperature of inversion and is denoted by T_i. Thus temperature of the gas will remain unchanged if $T_i = \frac{2\alpha}{Rb}$.

(iii) dT will be negative if $\left(\frac{2\alpha}{RT} - b\right)$ is $-ve$

i.e. $$\frac{2\alpha}{RT} < b \quad \text{or} \quad T > \frac{2\alpha}{Rb} \quad \text{or} \quad T > T_i.$$

Hence the gas will be heated if its initial temperature is more than the temperature of inversion. The temperature of inversion of hydrogen and helium are much below the room temperature (–80°C and – 240°C respectively) and hence at ordinary temperatures these gases show a heating effect. If, anyhow, these gases are cooled below their temperature of inversion and then passed through the porous plug, they will also show a cooling effect.

***Distinction Between Joule Expansion, Joule-Thomson Expansion and Adiabatic Expansion*:** In all these three processes the system under consideration is isolated thermally from external systems i.e., $dQ = 0$ for all the three process. Hence according to the definition, all the three processes are adiabatic. But in Joule expansion, the system is also isolated mechanically from the external system and hence the net external work done by or on the gas is zero i.e., $W = 0$. In Joule Thomson expansion, on the contrary, the system is not isolated mechanically from the external systems and both the pistons can move keeping the pressures P_1 and P_2 constant. The net external work done by the gas in passing through the plug

$$W = P_2 V_2 - P_1 V_1.$$

In adiabatic expansion, the system is not mechanically isolated from the external systems and also there is no attempt to keep the pressure of the gas constant. The expanding gas performs external work against the pressure of the atmosphere which is given by

$$W = \int P . dV.$$

The energy required for doing this work against the constant external pressure is drawn from the internal energy of the gas which therefore decreases resulting in a fall of temperature of the gas.

Applying first law of thermodynamics $dQ = dU + dW$ to all these three processes, we have (since $dQ = 0$)

$$dU = 0 \rightarrow \text{Joule expansion}$$

$$dU = P_1 V_1 - P_2 V_2 \rightarrow \text{Joule Thomson expansion.}$$

$$dU = -\int P dV \rightarrow \text{adiabatic expansion.}$$

In the third case (adiabatic expansion), the external work is drawn from the internal energy of the gas itself and hence there will always be a large fall of temperature. Thus in adiabatic expansion cooling is mainly due to external work. In the second case (Joule Thomson expansion) the contribution of the net external work may be a cooling effect or a heating effect depending upon whether $P_2 V_2$ is greater or small than $P_1 V_1$. However in, Joule Thomson expansion, the cooling is mainly produced due to the internal work done by the gas against molecular attractions. In Joule's expansion where $dU = 0$, a real gas will always show some cooling effect.

PROBLEMS

1. *Calculate the fall in temperature produced by adiabatic throttling process of a gm. mole of oxygen when the pressure is reduced by 50 atmospheres. The initial temperature of the gas is 27°C. Given for oxygen Cp = 7.03 cal /mole. K, a = 1.32 litre² × atoms mole⁻², b = 3.12 × 10⁻² litre mole⁻¹ and R = 8.3 × Joule/K mole.*

Solution: The fall in temperature is given by

$$dT = \frac{P_1 - P_2}{C_p}\left(\frac{2\alpha}{RT} - b\right)$$

Here $P_1 - P_2 = 50$ atmosphere $= 50 \times 10^5\ \text{Nm}^{-2}$

$a = 1.32$ liter$^2 \times$ atoms mole$^{-2} = 0.132\ \text{Nm}^4\ \text{mole}^{-2}$

$b = 3.12 \times 10^{-2}$ liter mole$^{-1} = 3.12 \times 10^{-5}\ \text{m}^3\ \text{mole}^{-1}$

$T = 27°\text{C} = 300$ K and $C_p = 7.03$ cals. mole^{-1}

or $\quad Cp = 7.03 \times 4.2$ Joule mole^{-1}

$$\therefore \quad dT = \frac{50 \times 10^5}{7.03 \times 4.2}\left(\frac{2 \times 0.132}{8.3 \times 300} - 3.12 \times 10^{-5}\right)$$

$$= \frac{50 \times 10^5 \times 7.48 \times 10^{-5}}{7.03 \times 4.2} = 12.7°\ \text{C}.$$

2. *Calculate the change in temperature when helium gas suffers Joule-Thomson expansion at – 173°C, the pressure difference on the two sides of the plug being 20 atmospheres.*

Does the gas show a heating effect or a cooling effect in this expansion at – 173°C. Given R = 8.3 Joule /K mole and for helium Vander Waal's constant a = 0.0341 litre² atoms. mole⁻² and b = 0.0237 litre mole⁻¹.

Solution: The fall in temperature in Joule Thomson expansion is given by

$$dT = \frac{1}{C_p}\left(\frac{2a}{RT} - b\right) dP$$

Here dP = 20 atmosphere = 20×10^5 N/m²

T = – 173°C = 100K

a = 0.0341 liter² atomos. mole^{-2}

= 0.0341×10^{-1} Nm⁴ mole^{-2}

b = 0.0237 liter mole^{-1} = 0.0237×10^{-3} m³ mole^{-1}, and for a monoatomic gas like helium

$$Cp = \frac{5}{2} R = \frac{5}{2} \times 8.3 \text{ Joule/mole K.}$$

Hence fall in temperature

$$dT = \frac{20\times10^5}{5\times4.15}\left[\frac{2\times0.0341\times10^{-1}}{8.3\times100} - 0.0237\times10^{-3}\right]$$

$$= -1.5 \text{ K.}$$

Thus there is a rise in a temperature equal to 1.5 K.

Note : As – 173°C is above the temperature of inversion of helium (–240°C), it will show a heating effect at this temperature.

3. *The Vander Waal's constant for hydrogen are a = 0.247 atoms litre ² mole⁻² ; b = 2.65 × 10⁻² litre mole⁻¹ Calculate*

(i) the temperature of inversion

(ii) The Joule Thomson cooling for 2 atmosphere fall of pressure, initial temperature being 100K

(Given $R = \frac{22.4}{273}$ ***atoms litre mole^{-1} K^{-1} and one atmosphere pressure = 1.0 × 10^5 N/m^{2}J***

Solution: (i) The temperature of inversion is given by $T_i = \frac{2\alpha}{Rb}$.

Here $a = 0.247$ atoms liter^{-1} mole^{-2}

$= 0.247 \times 10^{-1}$ N/m^4 mole^{-2}

$b = 2.65 \times 10^{-2}$ liter mole^{-1} $= 2.65 \times 10^{-5}$ m^3 mole^{-1}

and $R = \frac{22.4}{273}$ atoms liter mole^{-1} K^{-1}

or $R = \frac{22.4}{273} \times 10^5 \times 10^{-3}$ N/m mole^{-1} K^{-1}

$= 8.2$ Joule mole^{-1} K^{-1}.

Hence $$T_i = \frac{2 \times 0.247 \times 10^{-1}}{8.2 \times 2.65 \times 10^{-5}}$$

$= 227.4$ K.

(ii) The Joule Thomson cooling is given by

$$dT = \frac{1}{C_p}\left(\frac{2\alpha}{RT} - b\right)dP$$

Here $dP = 2$ atoms $= 2 \times 10^5$ N/m^2.

$T = 100$ K

and for a diatomic gas like hydrogen

$$Cp = \frac{7}{2} \times R = \frac{7}{2} \times 8.2 \text{ Joule mole}^{-1} \text{K}^{-1}$$

$$\therefore \quad dT = \frac{2 \times 10^5}{7 \times 4.1}\left[\frac{2 \times 0.247 \times 10^{-1}}{8.2 \times 100} - 2.65 \times 10^{-5}\right]$$

$$= \frac{2 \times 10^5}{28.7}(3.374 \times 10^{-5}) = 0.24 \text{ K}.$$

4

Heat Flow

There are three ways by which heat may travel from one place to another; conduction, convection, and radiation. In the process of *conduction, heat flows from one particle of the body to another in the direction of lower temperature without any visible movement of the particles.* When one end of a metallic rod is placed in a fire and left there for some time, heat is conducted through the rod to the other end which gradually becomes warmer.

According to kinetic theory, the molecules of a body are in a state of unceasing oscillatory motion about their fixed positions of equilibrium. The molecules at the heated end absorb energy from the source of heat, which increases their molecular kinetic energy and as a result the molecules begin to vibrate about their mean position faster than before.

On colliding with their slower moving neighbours, they transfer a part of their energy to them, thus putting them in more quicker oscillations and hence they later show a rise of temperature. This collision and transfer of energy to the molecules in the next section goes on till the heat reaches the most distant parts of the rod. The process of the conduction may thus be compared to the vibration of a large number of bells which are linked together by some suitable lengths of the cords, so that when one bell starts ringing, the others also take up the motion and begin to ring.

Thermal Conduction

We thus find that in conduction the heat energy diffuses through a body by the action of molecules possessing greater kinetic energy to those having lesser. In solids, which are electric insulators, this action probably takes place by means of the elastic binding forces between the atoms, while in solid electrical conductors (metals) the purpose is served by the so-called free electrons present within the metal and behaving like the molecules of a gas. The heat energy is transferred from the hot to the cold end both by collisions and by the motion of electrons.

Variable and Steady States

Let a rod of uniform cross-section be heated at one end. Every cross-section of the rod receives heat by conduction from the adjacent cross-section towards the hot end. A portion of the total heat thus received is retained by the cross-section to increase its own temperature; another portion is transmitted from its surface to the surroundings and the rest is transmitted to the next cross-section to repeat the process.

The greater the distance of a cross-section from the heated end, the more heat is lost by radiation from the part of the rod between the heated end and the particular section and hence lower is its temperature. Thus there is gradual fall in temperature along the rod and since the temperature at various points of the rod is varying with time, it is said to be a variable state. In this state rate of temperature increase is proportional to $\frac{K}{\rho c}$ where K is the coefficient of thermal conductivity, ρ the density and c the specific heat of the material of the rod. $\frac{K}{\rho c}$ is called thermal diffusivity or thermometric conductivity.

For some time the temperature at various points on the rod goes on increasing till a state is reached when the temperature at each point attains a constant value. Such a state is called the stationary or steady state.

In this state the temperature of any point in the rod depends upon the coefficient of thermal conductivity. K is a function of its distance from the heated end and is independent of time, the heat reaching any section is transmitted to the next (except radiation losses) and no heat is retained by the section itself. The steady state is different from thermal equilibrium in which all parts of the rod must have the same temperature.

Isothermal Surfaces

We have seen above that conduction of heat is possible only when a difference of temperature is maintained between the different parts of a body and that the flow of heat takes place in the direction of fall of temperature. Thus when a rod is heated at one end, the heat is conducted through the rod to other end and the temperature along its length changes from point to point, but the temperature at all points of a normal cross-section is the same. If we take a cylinder whose ends are perpendicular to its geometrical axis and are maintained at different definite temperatures, then in steady state, planes normal to the axis are planes of equal temperatures.

Such surfaces at all points of which temperature is the same are known as isothermal surfaces. Thus the direction of flow of heat at any point in the rod is at right angles to the isothermal surface at that point.

Temperature Gradient

When one point of body is maintained at a higher temperature than the other, the flow of heat occurs from the point at higher temperature to that at lower temperature. The temperatures at the midway points decreases in the direction of flow of heat. The rate of change of temperature with distance in the direction of the heat flow, is known as the temperature gradient. Thus, if θ_1 and θ_2 are the temperatures of two points distance x apart, then the ratio $\frac{\theta_1 - \theta_2}{x}$ represents the temperature gradient between the points.

In case when heat escapes from the sides of the body, the temperature gradient is not uniform. In such a case, let θ and $\theta + \delta\theta$ be the temperatures at points x and $x + \delta x$ in the direction of heat flow. Then

$$\text{Temperature gradient} = \frac{\theta - (\theta + \delta\theta)}{\delta x} = -\frac{d\theta}{dx}.$$

In the limit → 0, it gives

$$\text{Temperature gradient} = -\frac{d\theta}{dx} \text{ (at the given point).}$$

The negative sign with $\frac{d\theta}{dx}$ indicates that the temperature decreases with increase in x or in the direction of flow of heat.

Fundamental Equation of Heat Conduction–The Coefficient of Thermal Conductivity : Consider a parallel faced slab of a solid material of thickness x and area of cross-section A with its left-hand face at higher uniform temperature q_1 and right-hand face at lower uniform temperature q_2, as shown in fig. If it is assumed that no heat is lost from the sides of the slab, the direction of heat flow will be from left to right and at all points normal to the faces of slab. Under steady state, the quantity of the heat Q, that flows from one side to the other depends:

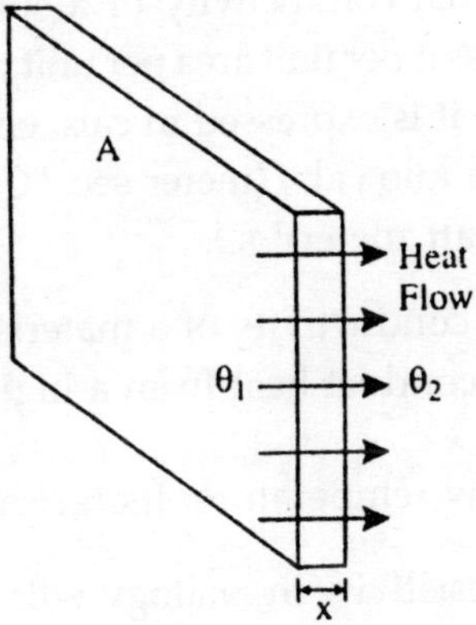

(i) directly on the face area A of the slab,

(ii) directly on the temperature gradient $\frac{\theta_1 - \theta_2}{x}$ across the slab,

(iii) directly on time t, for which the heat flows,

(iv) on the nature of the material of the slab,

Thus $$Q \propto \frac{A(\theta_1 - \theta_2)t}{x}$$

or $$Q = K.\frac{A(\theta_1 - \theta_2)t}{x}. \qquad ...(1)$$

where K is a constant whose value depends upon the nature of the material of the slab and is called the 'coefficient of thermal conductivity' or simply thermal conductivity of the material.

Equation (1) above is known as the fundamental equation of conduction of heat.

If in eqn. (1), we put

$$A = 1m^2, \frac{\theta_1 - \theta_2}{x} = 1^\circ / m. \text{ and } t = 1\,\text{sec}.,$$

then $Q = K.$

Hence 'thermal conductivity of a substance is the rate of flow of heat across it per unit area per unit temperature gradient'. In C.G.S. system it is expressed in cals. cm $^{-1}$ °C sec^{-1}. While in M.K.S. system in kilo cals/(meter sec. °C) or in Joule/(meter–sec. °C) or in watt meter^{-1} s^{-1}.

The thermal conductivity of a material represents the ease by which it can conduct heat from a high temperature region to a region of low temperature. Its reciprocal $\frac{1}{K}$ is, therefore, called thermal resistivity in analogy with electrical resistivity.

If we put in eqn. (1), $t = 1$, then rate of heat flow (which may be called heat current) is

$$q = \frac{Q}{t} = \frac{\theta_1 - \theta_2}{x/AK}.$$

The quantity $\frac{x}{AK}$ is called the thermal resistance.

Dimensions of K

$$Q = KA \cdot \frac{d\theta}{dx} \cdot t$$

$$\therefore \quad K = \frac{Q \cdot dx}{A \cdot d\theta \cdot t}$$

$$= \frac{[ML^2T^{-2}][L]}{[L^2][\theta][T]}$$

$$= [MLT^{-3}\theta^{-1}].$$

Thermal Diffusivity or Thermometric Conductivity : It is defined as the ratio of thermal conductivity to thermal capacity per unit volume and is denoted by h.

$$\therefore \quad h = \frac{K}{\left(\frac{mc}{V}\right)} = \frac{K}{\rho c}$$

where ρ is the density and c the specific heat of the material. It is a measure of rate of change of temperature (with time) when the body is not in steady state.

PROBLEMS

1. ***Calculate approximately the heat passing per hour through the walls and windows of a room (5 × 5 meters) if the walls are of bricks of thickness 30 cm and have windows of glass 3 m. m. thick and total area of 5 sq. meter. The temperature of the room is 30°C below that of the outside and the thermal conductivity of bricks and of glass is 12 × 10⁻⁴ and 25 × 10⁻⁴ C.G.S. units respectively.***

Solution : The area of the four walls of the room

$= 4 \times 5 \times 5 \times m^2 = 100\ m^2 = 100 \times 10^4\ cm^2.$

The area of the glass windows

$= 5\ m^2 = 5 \times 10^4 cm^2.$

∴ Area of the brick walls $= 100 \times 10^4 - 5 \times 10^4$

$= 95 \times 10^4$ cm.

Now the quantity of heat Q passing in time t is given by

$$Q = KA\frac{\theta_1 - \theta_2}{x}\,.\,t.$$

For brick walls, we have

$K = 12 \times 10^{-4}$ C.G.S. units,

$A = 95 \times 10^4\ cm^2,\ \theta_1 - \theta_2 = 30°C.$

$x = 30$ cm. and $t = 1$ hour = 3600 sec.

Hence the quantity of heat passing through the walls is

$$Q_1 = 12 \times 10^{-4} \times 95 \times 10^4 \times \frac{30}{30} \times 3600$$

$$= 4.104 \times 10^6 \text{ cal}.$$

Also, for glass windows, we have

$$K = 25 \times 10^{-4},\ A = 5 \times 10^4 \text{ cm}^2,\ \theta_1 - \theta_2 = 30°\text{C}.$$

$$x = 3 \text{ m.m.} = 0.3 \text{ cm. and } t = 3600 \text{ sec.}$$

Hence the quantity of heat passing through the windows is

$$Q_2 = 25 \times 10^{-4} \times 5 \times 10^4 \times \frac{30}{0.3} \times 3600$$

$$= 45 \times 10^6 \text{ cal}.$$

Therefore, the total amount of heat passing per hour into the room

$$= Q_1 + Q_2 = (4.104 \times 10^6 + 45 \times 10^6) \text{ cal.}$$

$$= 49.104 \times 10^6 \text{ cal.}$$

2. *Heat is supplied to a slab of cork, 5 cm. thick and of area 4 sq. meter by a heating coil spread over its surface. When the potential difference across the ends of the coil is 20 volt and current through it is 2.36 and the temperatures of the upper and lower faces are 12.5 and 0°C, calculate the thermal conductivity of the cork, if it may be assumed that whole of the heat developed in the coil is conducted through the slab.*

Solution : The heat developed in the coil is

Q = Current (in amp.) × potential diff. (in volt) Joule/sec.

= 2.36 × 20 Joule/sec.

= 47.2 Joule/sec.

This heat is conducted through the slab of the cork and hence from the relation

$$Q = KA.\frac{\theta_1 - \theta_2}{x} \qquad (\because\ t = 1 \text{ sec.})$$

Thermal conductivity

$$K = \frac{Qx}{A(\theta_1 - \theta_2)} = \frac{47.2 \times 0.5}{4 \times (12.5 - 0)}$$

$$= 4.72 \times 10^{-2} \text{ Joule/(meter sec. °C)}.$$

Heat Flow Through a Composite Slab

Consider a composite slab (or wall) made of two different materials A and *B* of the thickness x_1 and $x2$ and thermal conductivities K_1 and K_2 respectively. Let θ_1 and θ_2 be the temperatures of the end faces and θ the temperature at the common surface.

Let Q be the rate of flow of heat through the material A, then

$$Q = \frac{K_1 A(\theta_1 - \theta)}{x_1} \quad \text{... (1)}$$

Since there is no accumulation of heat anywhere in the steady state, the rate of heat flow through the material *B is* also the same, i.e.

$$Q = \frac{K_2 A(\theta - \theta_2)}{x_2} \quad \text{... (2)}$$

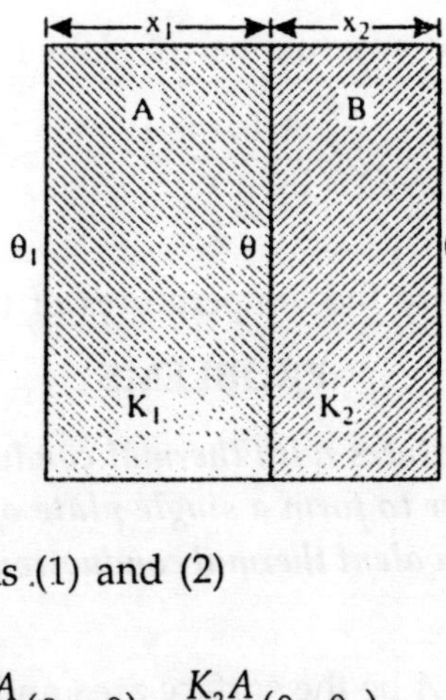

From equations (1) and (2)

$$\frac{K_1 A}{x_1}(\theta_1 - \theta) = \frac{K_2 A}{x_2}(\theta - \theta_2)$$

$$\frac{K_1}{x_1}\theta_1 + \frac{K_2}{x_2}\theta_2 = \left(\frac{K_1}{x_1} + \frac{K_2}{x_2}\right)$$

$$\therefore \quad \theta = \frac{\dfrac{K_1\theta_1}{x_1} + \dfrac{K_2\theta_2}{x_2}}{\dfrac{K_1}{x_1} + \dfrac{K_2}{x_2}}$$

Substituting the value of θ in eqn. (1), we have

$$Q = \frac{K_1 A}{x_1}\left[\theta_1 - \frac{K_1\theta_1 x_2 + K_2\theta_2 x_1}{K_1 x_2 + K_2 x_1}\right]$$

$$= \frac{K_1 A}{x_1}\left[\frac{K_2\theta_1 x_1 + K_2\theta_2 x_1}{K_1 x_2 + K_2 x_1}\right]$$

$$= \frac{K_1 A}{x_1}\left[\frac{K_2 x_1 + (\theta_1 - \theta_2)}{K_1 K_2\left(\dfrac{x_2}{K_2} + \dfrac{x_1}{K_1}\right)}\right]$$

$$= \frac{A(\theta_1 - \theta_n)}{\dfrac{x_1}{K_1} + \dfrac{x_2}{K_2}} \qquad \text{... (3)}$$

In general, for any number of slabs or walls (say *n*)

$$Q = \frac{A(\theta_1 - \theta_n)}{\Sigma \dfrac{x}{K}} \qquad \text{... (4)}$$

where θ_1 and θ_n are the temperatures of the extreme ends.

PROBLEMS

1. ***Two equal plates with thermal conductivities K_1 and K_2 are joined together to form a single plate of double thickness. Calculate the equivalent thermal conductivity of the composite plate so formed.***

Solution : Let *A* be the surface area and *x* the thickness of each plate. Also let q_1 and q_2 be the temperatures of the end faces and q the temperature of the surfaces in contact. Under steady conditions, the rate of heat flow through both the plates will be the same, hence

$$Q = \frac{K_1A(\theta_1 - \theta)}{x} = \frac{K_2A(\theta - \theta_2)}{x} \quad \text{... (1)}$$

or $$K_1\theta_1 - K_1\theta = K_2\theta - K_2\theta_2$$

or $$K_1\theta_1 + K_2\theta_2 = \theta\,(K_1 + K_2)$$

$\therefore$ $$\theta = \frac{K_1\theta_1 + K_2\theta_2}{K_1 + K_2} \quad \text{... (2)}$$

Now the thickness of composite plate is $2x$ and let K be its equivalent thermal conductivity, then rate of heat flow through it

$$Q = \frac{KA(\theta_1 - \theta_2)}{x} \quad \text{... (3)}$$

Equations (1) and (3) give

$$\frac{K_1A(\theta_1 - \theta)}{x} = \frac{KA(\theta_1 - \theta_2)}{x}$$

or $$K = \frac{2K_1\,(\theta_1 - \theta)}{(\theta_1 - \theta_2)}$$

or $$K = \frac{2K_1}{\theta_1 - \theta_2}\left[\theta_1 - \frac{K_1\theta_1 + K_2\theta_2}{K_1 + K_2}\right] \quad \text{... using (2)}$$

or $$= \frac{2K_1}{\theta_1 - \theta_2}\left[\frac{K_1\theta_1 + K_2\theta_1 + K_1\theta_1 + K_2\theta_2}{K_1 + K_2}\right]$$

$$= \frac{2K_1 + K_2}{K_1 + K_2}.$$

Aliter

$\therefore$ Thermal resistance $= \dfrac{x}{KA}$

$\therefore$ Thermal resistance of first plate $= \dfrac{x}{K_1A}$

Thermal resistance of second plate $= \dfrac{x}{K_2A}$

Thermal resistance of composite plate $= \dfrac{2x}{KA}$

where K *is* the equivalent thermal conductivity of composite plate. Since the plates can be supposed to be connected in series,

$$\therefore \quad \frac{2x}{KA} = \frac{x}{K_1 A} + \frac{x}{K_2 A}$$

$$\text{or} \quad \frac{2}{K} = \frac{1}{K_1} + \frac{x}{K_2} = \frac{K_1 + K_2}{K_1 K_2}$$

$$\therefore \quad K = \frac{2K_1 K_2}{K_1 + K_2}$$

2. *The walls of a chamber are made of insulating bricks 8 cm thick. On the inside the walls are lined with cork 4 cm thick. The inside temperature is 10°C while the outside temperature is 20°C. Determine the temperature of the interface. Given, K for bricks = 0.0004 and K for the cork = 0.0001. Determine also the temperature gradient in each section of the wall.*

Solution : Let q be the temperature of the interface. The rate of flow of heat in the steady state through the two sections of the wall (bricks and cork) will be the same. Hence

$$0.0004 \times A \times \frac{20-\theta}{8} = 0.0001 \times A \times \frac{\theta - 10}{4}$$

$$\text{or} \quad \frac{20-\theta}{2} = \frac{\theta - 10}{4}$$

$$\text{or} \quad = 16.6°\text{C}.$$

Now drop of temperature in bricks = 20 – 16.6 = 3.4°C.

Thickness of bricks = 8 cm.

$$\therefore \quad \text{temperature gradient in bricks} = \frac{3.4}{8} = 0.42°\text{C/cm}.$$

$$\text{Similarly temperature gradient in cork} = \frac{16.6-10}{4} = \frac{6.6}{4}$$

$$= 1.65°\text{C/cm}.$$

3. *A composite flat sheet is made up of 1.2 cm. thickness of one material of conductivity 3 × 10^{-3} cal. cm^{-1} deg^{-1} sec^{-1} and 0.8 cm thickness of another material of conductivity 0.06 × 10^{-3} cal. cm^{-1} deg^{-1} sec^{-1} joined together. If the open face of the first material is at temperature –10°C and open face of the second material is at temperature +30°C, calculate (i) the temperature of the interface between the two materials, (ii) the rate of transfer of heat across the sheet per m^2 per minute.*

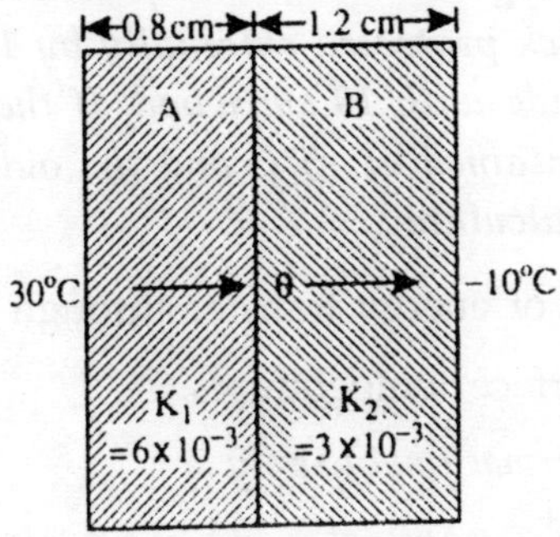

Solution : Let the two materials be *A* and *B* joined as shown in fig. Let temperature of the interface be θ.

Amount of heat flowing in time t through the material *A*

$$Q_1 = \frac{0.6\times10^{-3}\times A(30-\theta)}{0.8}\,.\,t \qquad \text{... (1)}$$

And the heat flowing in time t through the material *B*

$$Q_2 = \frac{3\times10^{-3}\times A[\theta-(-10)]}{1.2}.t = \frac{3\times10^{-3}\times A(\theta+10)}{1.2}.t \qquad \text{... (2)}$$

In the steady state, $Q_1 = Q_2 = Q$

or $$\frac{0.6\times10^{-3}\times A(30-\theta)}{0.8}\,t = \frac{3\times10^{-3}\times A(\theta+10)}{1.2}\,t.$$

on solving we get, θ = – 0.77°C.

Now the rate of transfer of heat across the sheet per square meter per minute may be obtained by putting $\theta = -0.77^\circ C$, $t = 60$ sec. and $A = 1\ m^2 = 10^4\ cm^2$ in eqn. (1). Hence

$$Q = Q_1 = \frac{0.6 \times 10^{-3} \times 10^{4}\left[30-(-0.77)\right]}{0.8} \times 60$$

$$= \frac{6 \times 30.77 \times 60}{0.8} = 13854 \text{ cal.}$$

4. *An insulating wall of area 4 square meter is made from cork 16.5 cm thick protected externally by 12.5 cm brick and lined on the inside with 10 cm wood. If the exposed surface of wood is maintained at –5°C and the outer surface of the brick at 20°C, calculate :*

(i) The rate of energy transfer through the wall, and

(ii) The interface temperatures.

Given thermal conductivity

For brick = 2.5×10^{-3}, *For cork* = 1.1×10^{-4} *and For wood* = 4×10^{-4} *C.G.S. units.*

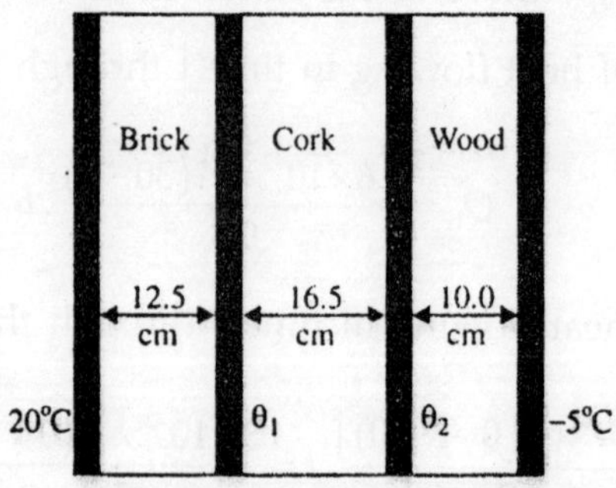

Solution : Let the surfaces, and their temperatures be as shown in fig. The rate of flow of heat through any surface is given by

$$Q = \frac{KA(\theta - \theta')}{x}$$

Here $A = 4 \times 10^4$ sq. cm. and for brick $K = 2.5 \times 10^{-3}$ cal. $sec^{-1}\ cm^{-1}\ ^\circ C^{-1}$.

Hence through bricks, rate of heat flow

$$Q_1 = \frac{2.5\times10^{-3}\times4\times10^{4}(20-\theta_1)}{12.5} = 8(20-\theta_1) \quad \text{... (1)}$$

Similarly, through cork and wood, the rate of heat flow are respectively

$$Q_2 = \frac{1.1\times10^{-4}\times4\times10^{4}[\theta_2-(-5)]}{16.5} = \frac{4}{15}(\theta_1-\theta_2) \quad \text{... (2)}$$

$$Q_3 = \frac{4\times10^{-4}\times4\times10^{4}\times[\theta_2-(-5)]}{10} = \frac{8}{5}(\theta_2+5) \quad \text{... (3)}$$

In the steady state, the rate of flow through all the surfaces is the same. Hence from (1) and (2)

$$8(20-\theta_1) = \frac{4}{15}(\theta_1-\theta_2)$$

or
$$31\theta_1-\theta_2 = 600 \quad \text{... (4)}$$

Similarly from (2) and (3)

$$\frac{4}{15}(\theta_1-\theta_2) = \frac{8}{5}(\theta_2+5)$$

or
$$\theta_1-7\theta_2 = 30 \quad \text{... (5)}$$

Solving (4) and (5),

$$\theta_1 = 19.5°\text{C and } \theta_2 = -1.5°\text{C}.$$

Hence the interface temperatures are 19.5°C, and – 1.5°C. Substituting for 02 in equation (3)

$$Q_3 = \frac{8}{5}(-1.5+5) = \frac{8}{5}\times3.5$$

$$= 5.5 \text{ cm/sec.}$$

Hence the rate of energy transfers through the wood (or, wall as whole) is 5.5 cal/sec.

5. ***A bar of length 40 cm and uniform cross section 5 cm^2 consists of two halves, AB of copper and BC of iron welded***

together at B. The end A is maintained at 200°C and end C at 0°C and the sides are thermally insulated. Find the rate of flow of heat along the bar when the steady state has been reached. (Thermal conductivities are : copper 0.9; iron 0.12).

Solution : Under steady conditions, the rate of heat flow through both sections of the bar will be same i.e.,

$$\frac{Q}{t} = \frac{K_1 A(200-\theta)}{d} = \frac{K_2 A(\theta - 0)}{d} \qquad \text{... (1)}$$

or $$0.9 \times (200-\theta) = 0.12(\theta - 0)$$

or $$\theta = 176.4°\text{C}.$$

Putting the value, $\theta = 176.4°$C, $A = 5$ cm^2 and $d = \frac{40}{2} = 20$ in eqn. (1), rate of flow of heat along the bar

$$\frac{Q}{t} = \frac{0.9 \times 5 \times (200-176.4)}{20}$$

$$= 5.31 \text{ cals/sec.}$$

6. Three slabs of 20 sq. cm cross section and 2 cm thick are piled one on top of the other. Their thermal conductivities are 0.1, 0.2 and 0.4 C.G.S. units. Find the thermal conductivity of a single slab 5 sq. cm in cross-section and 2.5 cm thick which conducts the same heat per sec. with the same temperature difference between its faces.

Solution : Referring to eqn. (4) article 9.6, the rate of heat flow through a number of slabs is given by

$$q = \frac{Q}{t} = \frac{A(\theta_1 - \theta_n)}{\Sigma \frac{x}{K}}$$

$$\Sigma \frac{x}{K} = \frac{x_1}{K_1} + \frac{x_2}{K_2} + \frac{x_3}{K_3} = \frac{2}{0.1} + \frac{2}{0.2} + \frac{2}{0.4} = 35.$$

$$\therefore \quad q = \frac{20(\theta_1 - \theta_2)}{35} \qquad (\because A = 20 \text{ cm}^2 \text{ given})$$

But this is equal to the heat flow through a single slab of thickness 2.5 cm and 5 cm² in cross-section of conductivity K (say) for the same temperature difference $(\theta_1 - \theta_2)$,

$$\therefore \quad q = \frac{K \times 5 \times (\theta_1 - \theta_2)}{2.5}$$

$$\therefore \quad = \frac{20(\theta_1 - \theta_2)}{35} = \frac{K \times 5 \times (\theta_1 - \theta_2)}{2.5}$$

or

$$= K\frac{20 \times 2.5}{35 \times 5} = \frac{1}{3.5}$$

$$= 0.286 \text{ C.G.S. units}$$

Growth of Ice on Ponds

The cold air above the water in a pond abstracts heat from the water. As a result, the water begins to freeze into ice layer. In order to calculate the rate of growth of this layer, let us consider that a x cm. Thick layer of ice has already formed on the surface of the pond. Let θ°C be temperature of water below ice and –θ°C the temperature of air above the surface of ice. Let a further layer of ice of thickness dx be formed in time dt. Then, if A is the surface area of the layer and ρ the density of ice, then mass of ice formed = $A\,dxp$. If L is the latent heat of fusion of ice, then heat given up by water = $A\,dxp \times L$ calories. (i)

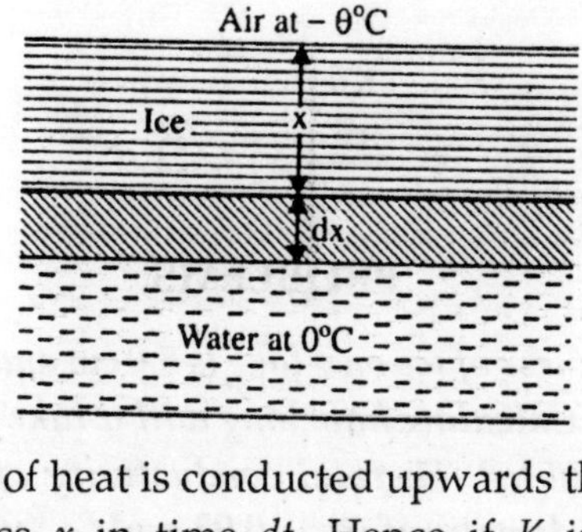

This amount of heat is conducted upwards through the ice layer of thickness x in time dt. Hence if K is the thermal conductivity of ice, heat conducted upwards

$$KA.\frac{0-(-\theta)}{x}dt = KA.\frac{\theta}{x}dt. \quad \text{....(ii)}$$

Equating (i) and (ii)

$$KA\frac{\theta}{x}dt = A\,dx\,\rho L$$

or the rate of growth of ice layer $\frac{dx}{dt} = \frac{K\theta}{x\,\rho L}$ ... (iii)

$$\therefore \quad dt = \frac{x\,\rho L}{K\theta}dx.$$

Therefore, the total time in which the thickness of ice will increase by x

$$\int dt = \frac{\rho L}{K\theta}\int x dx \quad \text{... (iv)}$$

or

$$t = \frac{\rho L}{K\theta}.\frac{x^2}{2} + \text{constant.}$$

If at $t = 0$, $x = 0$, the constant comes out to be zero.

Then

$$t = \frac{\rho L}{2K\theta}x^2. \quad \text{... (v)}$$

Time taken by the ice layer to increase in thickness from x_1 to x_2

$$\int_0^t dt = \frac{\rho L}{K\theta}\int_{x1}^{x2} x dx$$

or

$$t = \frac{\rho L}{2K\theta}\left(x_2^2 - x_1^2\right) \quad \text{... (v)}$$

PROBLEMS

1. *The thickness of ice on a lake is 5.0 cm and the temperature of air is –20°C. Calculate how long will it take for the thickness of ice to be doubled. Thermal conductivity of ice = 0.005 cal. cm^{-1} sec^{-1} °C^{-1}, density of ice = 0.92 gmlc.c. and latent heat of ice = 80 cal /gm.*

Solution : The time for the thickness of ice layer to increase by dx is given by

$$dt = \frac{x\rho L}{K\theta} dx.$$

Therefore the time in which the thickness will increase from x_1 to x_2

$$t = \frac{\rho L}{K\theta} \int_{x_1}^{x_2} x dx$$

or
$$t = \frac{\rho L}{K\theta} \left(\frac{x^2}{2} \right)_{x_1}^{x_2} = \frac{\rho L}{2K\theta} \left(x_2^2 - x_1^2 \right).$$

Here $\rho = 0.92$ gm/c.c., $L = 80$ cal/gm, $K = 0.005$ C.G.S. units $-\theta = -20°C$ or $\theta = 20°C$, $x_1 = 5$ cm, $x_2 = 10$ cm (doubled)

$$t = \frac{0.92 \times 80}{2 \times 0.005 \times 20} \left(10^2 - 5^2 \right)$$

$= 27600$ sec.

$= 7$ hour 40 minutes.

2. ***If the thermal conductivity of ice = 0.005, its latent heat = 80 cal /gm and density at 0°C = 0.91, find the time in which a layer of ice 3 cm thick on the surface of the pond will increase in thickness by 1 mm, when the temperature of surrounding air is 20°C below zero.***

Solution : The time for the thickness of ice layer to increase from x_1 to x_2

$$t = \frac{\rho L}{K\theta} \int_{x_1}^{x_2} x dx$$

or
$$= \frac{\rho L}{2K\theta} \left(x_2^2 - x_1^2 \right).$$

Given, $\rho = 0.91$ gm/c.c., $L = 80$ cal/gm, $K = 0.005$

$-\theta = -20°C$ or $\theta = 20°C$, $x_1 = 3$ cm, $x_2 = 3 + 0.1 = 3.1$ cm.

$$\therefore \quad t = \frac{0.92 \times 80}{2 \times 0.005 \times 20}\left[(3.1)^2 - (3)^2\right]$$

$$\approx 222 \text{ sec.} = 3 \text{ min. } 42 \text{ sec.}$$

3. ***How much time will it take to increase the thickness of ice in a pond from 20 cm. to 30 cm ?***

***Temperature of atmosphere* = – 15°C**

***K* = *0.005 C.G.S. units, L* = *80 cal /gm, p* = *0.9 gm./cm*³.**

Solution : Time taken in increasing the ice thickness from x_1 to x_2

$$t = \frac{\rho L}{2K\theta}\left(x_2^2 - x_1^2\right)$$

Here, $x_1 = 20$ cm, $x_2 = 30$ cm.

Substituting the given values

$$t = \frac{0.9 \times 80}{2 \times 0.005 \times 15}\left[(30)^2 - (20)^2\right]$$

$$= 24 \times 10^4 \text{ sec.}$$

$$= 66 \text{ hr. } 40 \text{ min.}$$

Temperature Distribution in a Rod Heated at One End (Rectilinear Flow of Heat along a Metal Bar)

Fourier Equation of Heat Flow : Let us consider a long metal bar of uniform area of cross-section placed parallel to x-axis and heated at one end. Let hot end be taken as the origin of coordinate system (Fig.).

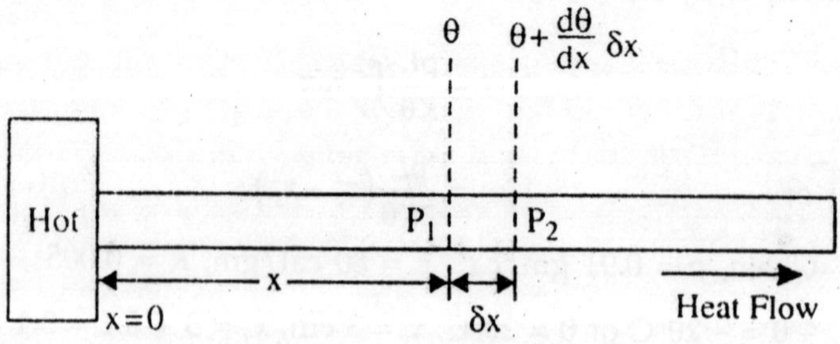

Let us consider two parallel planes P_1 and P_2 perpendicular to the length of the bar at distance x and $x + \delta x$ from the hot end. Let θ be the excess of temperature above surroundings at the plane P_1 and $\frac{d\theta}{dx}$ the temperature gradient. Then

Excess of temperature at $P_2 = \theta + \frac{d\theta}{dx}, \delta x$

and temperature gradient at $P_2 = \frac{d}{dx}\left(\theta + \frac{d\theta}{dx}.\delta x\right)$

If Q_1 and Q_2 are respectively the quantities of heat entering the plane at P_1 and leaving the plane at P_2 per second, then

$$Q_1 = -KA\frac{d\theta}{dx}$$

where A is the cross-sectional area of the bar and K the thermal conductivity of its material, and

$$Q_2 = -KA\frac{d}{dx}\left(\theta + \frac{d\theta}{dx}\delta x\right).$$

Therefore net gain of heat per second by the section δx between P_1 and P_2 of the rod

$$Q = Q_1 - Q_2 = -KA\frac{d\theta}{dx} - \left\{-KA\frac{d}{dx}\left(\theta + \frac{d\theta}{dx}\delta x\right)\right\}$$

$$= KA\frac{d^2\theta}{dx^2}\delta x. \qquad \text{... (i)}$$

CASE I :

Before the State is Reached : The amount of heat Q is partly used up in raising the temperature of the section and the rest is lost to the surroundings due to radiations. Let ρ be the density, c the specific heat of the material of bar, and $\frac{d\theta}{dt}$ be the rate of rise of temperature, then amount of heat used up in raising the temperature of the section.

= mass of the section × specific heat × rise in temp. per sec.

$$=(A\delta x)\rho \times c \times \frac{d\theta}{dt}. \quad \text{... (ii)}$$

Also, if E is the emissivity of the surface, i.e., the heat lost per second per unit area of the surface per unit temperature excess over the surroundings, and p the perimeter of the cross-section of the bar, then heat lost per second due to radiation

= emissivity × surface area of the section × temp excess

$$= E \times p.\delta x \times \theta. \quad \text{... (iii)}$$

Hence from (i), (ii) and (iii)

$$Q = KA\frac{d^2\theta}{dx^2}\delta x$$

$$= A\,\delta x\,\rho \times c \times \frac{d\theta}{dt} + E \times \rho\,\delta x \times \theta. \quad \text{... (iv)}$$

Dividing throughout by $KA\ \delta x$, we get

$$\frac{d^2\theta}{dx^2} = \frac{\rho c}{K}.\frac{d\theta}{dt} + \frac{E_p}{KA} \times \theta.$$

This is the general equation representing the unidirectional rectilinear flow of heat and is known as Fourier's differential equation.

If the heat lost by radiations is negligible, i.e., the bar is covered or the bar is perfectly insulated from surroundings, then its emissivity $E = 0$ and eqn. (iv) reduces to

$$\frac{d^2\theta}{dx^2} = \frac{\rho c}{K}.\frac{d\theta}{dt}$$

or

$$\frac{d\theta}{dt} = \frac{K}{\rho c}.\frac{d^2\theta}{dx^2} = h\frac{d^2\theta}{dx^2} \quad \text{... (v)}$$

where $\frac{K}{\rho c} = h$ is the thermal diffusivity, which determines the rate at which temperature changes take place in the bar.

Case II.

After the Steady State is Reached : The temperature at every point of the bar becomes stationary, i.e., $\frac{d\theta}{dt} = 0$. Equation (iv) then reduces to

for an exposed bar $$\frac{d^2\theta}{dx^2} = \frac{E_P}{KA}.\theta = \mu^2\theta \qquad \text{... (vi)}$$

where $\frac{E_P}{KA} = \mu^2$.

In order to solve this differential equation, let us put

$$\theta = \alpha e^{\alpha x} \qquad \text{... (vii)}$$

where a and α are constants. Differentiating it twice, we have

$$\frac{d\theta}{dt} = a\ \alpha e^{\alpha x} \text{ and } \frac{d^2\theta}{dx^2}\alpha^2.\alpha e^{\alpha x}$$

$$\therefore \qquad \alpha^2 a e^{\alpha x} = \mu^2 . \alpha e^{\alpha x}$$

or $$\alpha^2 = \mu^2$$

or $$\alpha = \pm\mu.$$

Substituting in eqn. (vii), the value of α, we get the solution of eqn. (vi) to be

$$\theta = \alpha\ e^{=\mu x}$$

Hence the most general solution can be written as

$$\theta = \alpha_1 e^{=\mu x} + \alpha_2 e^{-\mu x} \qquad \text{... (viii)}$$

If the bar is supposed to be infinitely long, its cold end will be nearly at the temperature of the surroundings, i.e., at $x = \infty$, $\theta = 0$. Also if θ_0 is the excess temperature of the hot end above surroundings, then at $x = 0$, $\theta = \theta_0$. Applying the first condition, we have from equation (viii)

$$\theta = \alpha_1 e^{+\mu x} + \alpha_1 e^{-\mu\infty}$$

or $$0 = \alpha_1 e^{+\mu\infty} \qquad (\because\ e^{-\mu\infty} = 0)$$

$$\therefore \qquad \alpha_1 = 0 \qquad (\because\ e^{-\mu 0} = 1)$$

Applying the second condition (at $x = 0, \theta = \theta_0$) and putting $\alpha_l = 0$ in eqn. (viii), we get

$$\theta_0 = 0 + a_2 e^{-\mu 0}$$

or $$\alpha_2 = \theta_0. \qquad (\because\ e^{-\mu 0} = 1)$$

Substituting the values of α_1 and α_2 in eq. (viii), it becomes

$$\theta = \theta_0\, e^{-\mu x}, \qquad \text{... (ix)}$$

This equation gives the excess of temperature of a point at a distance x from the hot end along the bar exposed to radiations after the steady state has been reached and represents that excess temperature is an exponential function of distance x and is independent of time. It has been shown graphically in the below given figure.

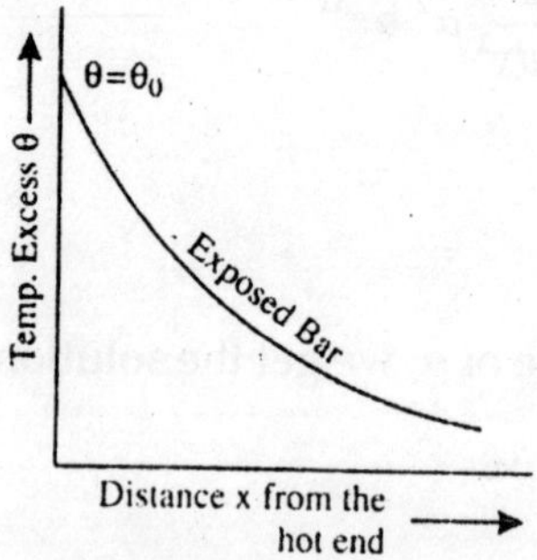

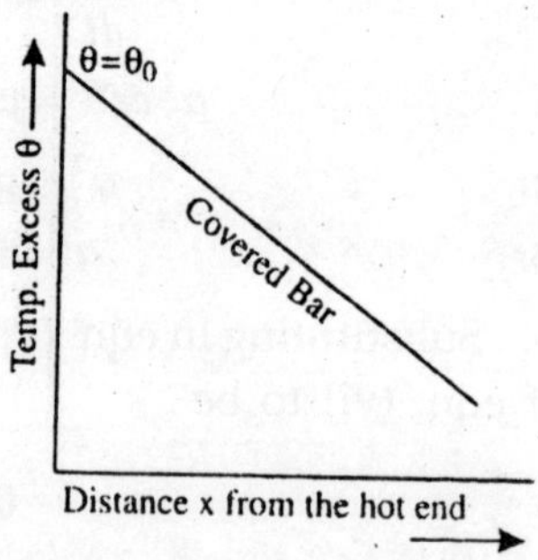

If the radiations loss is negligible i.e., the bar is covered, Then $E = 0$ and hence

$$\mu^2 = \frac{E_p}{KA} = 0.$$

Equation (iv), then gives

$$\frac{d^2\theta}{dx^2} = 0.$$

Integrating it,

$$\frac{d\theta}{dt} = \alpha'_1\,(\text{constant}).$$

Again integrating

$$\theta = \alpha_1' x + \alpha_2' \text{ (constnat)} \qquad (x)$$

To evaluate constants a_1' and a_2', let l be the length of the bar and its cold end be nearly at the temperature of the surroundings. Then the boundary conditions are,

At $x = l, \theta = 0$

and at $x = 0, \theta = \theta_0,$

Substituting these conditions in eqn. (x), we get

$$0 = a_1'l + a_2'$$

and $\theta_0 = 0 + a_2'$

$\therefore$ $a_1'l + \theta_0 = 0$

or $a_1' = -\dfrac{\theta_0}{l}$

Putting $\alpha'_1 = \frac{\theta_0}{l}$ and $\alpha'_2 = \theta_0$ we have

$$= \theta_0 = \frac{-\theta_0}{l} x + \theta_0 = \theta_0 - \frac{-\theta_0}{l} x. \qquad \text{... (xii)}$$

This equation gives the excess temperature at any point at a distance x from the heated end, when the bar is covered and hence no radiations taking place. From this, it is obvious that the temperature falls off linearly from the hot end, as shown in the figure.

Ingen Hausz Experiment

It is used to compare the thermal conductivities of different materials. For it, materials are taken in the form of long thin rods, identical in size and shape and polished similarly. They are then coated identically with wax and fixed with one end inserted in an oil bath (Fig.) in which oil is maintained at its boiling point.

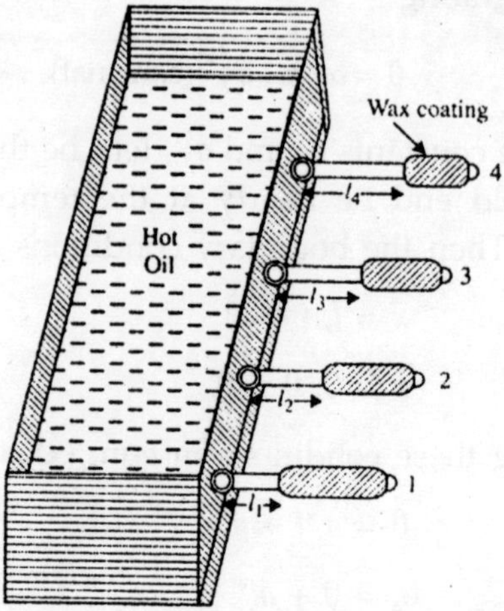

Heat is conducted along the rods and in steady state wax is found to be melted to different lengths depending upon the value of K for different materials.

Let l_1, l_2, l_3 ...... be the lengths of wax melted along the bars 1, 2, 3 ...... and K_1, K_2, K_3 ..... be their thermal conductivities, then

$$\frac{K_1}{l_1^2} = \frac{K_2}{l_2^2} = \frac{K_3}{l_3^2} = \ldots$$

from which the values of K can be compared.

Proof : Let the rods be sufficiently long so that their free ends be taken at the temperature of the surroundings. Let θ_0 be the excess temperature of the heated ends, then the equation representing the temperature distribution along each bar in the experiment is $\theta_0 = \theta_0 e^{-\mu x}$ [Refer eqn. (ix) article 9.8].

Let θ_m be the melting point of wax, then $\theta = \theta_m$, at $x = l_1$ for bar 1, at $x = l_2$ for bar 2, at $x = l_3$ for bar 3.

$\therefore$ For bar 1, $\theta_m = \theta_0\, e^{-\mu_1 l_1}$

For bar 2, $\theta_m = \theta_0\, e^{-\mu_2 l_2}$

and For bar 3, $\theta_m = \theta_0\, e^{-\mu_3 l_3}$

Hence $\mu_1 l_1 = \mu_2 l_2 = \mu_3 l_3$

or $\mu_1^2 l_1^2 = \mu_2^2 l_2^2 = \mu_3^2 l_3^2$.

But $\mu_1^2 = \frac{Ep}{K_1 A}, \mu_2^2 = \frac{Ep}{K_2 A}$ and $\mu_3^2 = \frac{Ep}{K_3 A}$; because *E, p* and *A* for all the rods are the same, as they are identical in size, shape, etc.

$$\therefore \quad \frac{Ep}{K_1 A}, l_1^2 = \frac{Ep}{K_2 A} \ l_2^2 = \frac{Ep}{K_3 A} l_3^2$$

or
$$\frac{K_1}{l_1^2} = \frac{K_2}{l_2^2} = \frac{K_3}{l_3^2}$$

or
$$K_1 : K_2 : K_3 = l_1^2 : l_2^2 : l_3^2$$

i.e., the conductivities of the different rods vary as the square of the lengths upto which wax has melted on them.

Temperature Distribution in a Cylinder Heated Along The Axis (Cylindrical Flow of Heat) : Let us consider a cylindrical tube of length l, inner radius *r*1 and outer radius *r*2. Let the tube be heated along its axis by placing an electrically-heated wire. Heat will flow radially from the inner side towards the outer side across the walls of the tube. When the steady state is reached, let the constant temperature of the inner and outer surface of the tube be q_1 and q_2 respectively ($q_1 > q_2$).

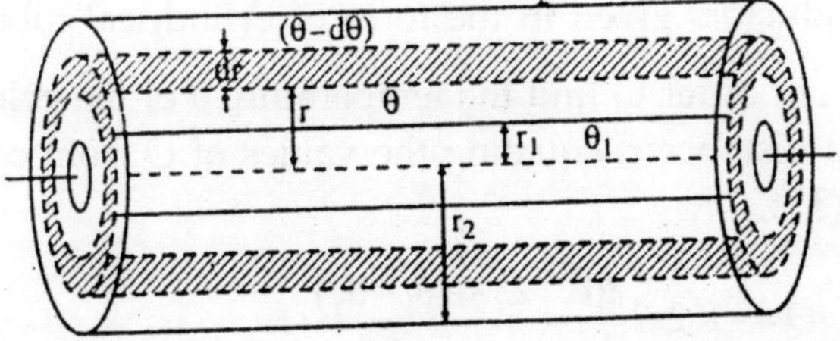

Consider a thin cylindrical shell of thickness *dr* at a distance *r* from the axis (Fig.). Let θ and (θ – *d*θ) be the temperatures at *r* and (*r* + *dr*). The fall in temperature on travelling outwards by dr is $d\theta$, hence the temperature gradient is $-\frac{d\theta}{dr}$. The amount of heat flowing per second through the shell of radius r (and hence surface area $2\pi rl$) is given by

$$Q = -K\,2\pi rl\,\frac{d\theta}{dr} \qquad \text{... (1)}$$

where K is the thermal conductivity. The negative sign expresses the fact that θ diminishes with an increase in r.

Rearranging eqn. (1), we get

$$Q\frac{dr}{r} = -2\pi rK\,l\,d\theta. \qquad \text{... (2)}$$

After a steady state is reached, Q becomes constant. Integrating both sides of eqn. (2) between their respective limits, we get

$$Q\int_{r_1}^{r_2}\frac{dr}{r} = -2\pi Kl\int_{\theta_2}^{\theta_2} d\theta$$

or $$Q\log_e\frac{r_2}{r_1} = 2\pi Kl(\theta_1 - \theta_2) \qquad \text{... (3)}$$

$\therefore$ $$K = \frac{Q\log_e\frac{r_2}{r_1}}{2\pi l(\theta_1 - \theta_2)}. \qquad \text{... (4)}$$

This expression is used to find the thermal conductivity of poor conductors given in the form of cylindrical tubes.

Now, in order to find the temperature θ of the cylindrical shell at a distance r, equating the values of Q from equation (1) and (3) :

$$Q = -K.2\pi l\frac{d\theta}{dr} = \frac{2\pi Kl(\theta_1 - \theta_2)}{\log_e\frac{r_2}{r_1}}$$

or $$d\theta = -\frac{(\theta_1 - \theta_2)}{\log_e\frac{r_2}{r_1}}.\frac{dr}{r}.$$

Integrating it, we get

$$\theta = -\frac{(\theta_1 - \theta_2)}{\log_e \frac{r_2}{r_1}} \log_e r + C \text{(constant)}. \qquad ...(5)$$

But when $r = r_1$, $\theta = \theta_1$

$$\theta_1 = -\frac{(\theta_1 - \theta_2)}{\log_e \frac{r_2}{r_1}} \log_e r_1 + C$$

or

$$C = \theta_1 + \frac{(\theta_1 - \theta_2)}{\log_e \frac{r_2}{r_1}} \log_e r_1.$$

Substituting the value of C in eqn. (5), we get

$$\theta = -\frac{(\theta_1 - \theta_2)}{\log_e \frac{r_2}{r_1}} \log_e r + \theta_1 + \frac{(\theta_1 - \theta_2)}{\log_e \frac{r_2}{r_1}} \log_e r_1$$

$$= \frac{1}{\log_e \frac{r_2}{r_1}} \left[-(\theta_1 - \theta_2) \log_e r + \theta_1 \log_e \frac{r_2}{r_1} + (\theta_1 - \theta_2) \log_e r_1 \right]$$

$$= \frac{1}{\log_e \frac{r_2}{r_1}} [-(\theta_1 - \theta_2) \log_e r + \theta_1 \log_e r_2 - \theta_1 \log_e r_1 + \theta_1 \log_e r_1 - \theta_2 \log_e r_1]$$

$$= \frac{1}{\log_e \frac{r_2}{r_1}} \left[(\theta_1 \log_e r_2 - \theta_2 \log_e r_1) - (\theta_1 - \theta_2) \log_e r \right] \qquad ...(6)$$

This equation represents the temperature distribution in a cylindrical tube heated along the axis in steady state.

PROBLEMS

1. *Along the axis of a long uniform cylindrical shell of inner and outer radii 4 and 12 cm, there is a wire in which 7.0 watt of energy is developed per cm. length. If the conductivity*

of the material is* 0.003 *joule* cm^{-1} deg^{-1} sec^{-1} *and the steady state temperature of outer surface is* 20°C, *deduce (i) the temperature at the inner face of the cylinder, (ii) the temperature at* $r = 8$ *m.

Solution : (i) The energy developed in the wire = 7 watt/cm.

$$\therefore \qquad \frac{Q}{l} = 7 \text{ joules/cm. sec.}$$

$$= \frac{7}{J} = \frac{7}{4.2} = \frac{5}{3} \text{ cals/cm.sec.}$$

Also, the conductivity of material

$$K = 0.003 \text{ joule cm}^{-1} \text{ deg}^{-1} \text{ sec}^{-1}$$

$$= \frac{0.003}{4.2} = \frac{0.01}{14} \text{cals. cm}^{-1}\text{sec}^{-1}$$

Now the conductivity of a cylindrical tube is given by

$$K = \frac{Q \log_e \frac{r_2}{r_1}}{2\pi l(\theta_1 - \theta_2)} = \frac{Q \times 2.3 \log_{10} \frac{r_2}{r_1}}{2\pi l(\theta_1 - \theta_2)} \; (\because \; \log_e x = 2.3 \; \log_{10} x)$$

Here, given $r_1 = 4$ cm, $r_2 = 12$ cm, $\theta_2 = 20°$C, $\theta = ?$

$$K = \frac{0.01}{14} = \frac{5 \times 2.3 \times \log_{10} \frac{12}{4}}{3 \times 2 \times 3.14 \times (\theta_1 - \theta_2)} \qquad \left[\because \frac{Q}{l} = \frac{5}{3}\right]$$

$$\text{or} \quad (\theta_1 - \theta_2) = \frac{5 \times 2.3 \times \log_{10} 3 \times 14}{3 \times 2 \times 3.14 \times 0.01}$$

$$= \frac{5 \times 2.3 \times 0.477 \times 14}{3 \times 2 \times 3.14 \times 0.01} = 407.7$$

$$\therefore \quad \theta_1 = 407.7 + \theta_2 = 407.7 + 20 = 427.7°\text{C}.$$

(ii) Now the temperature of a cylindrical shell at a distance r = 8 cm from the axis

$$\theta = \frac{1}{\log_{10}\frac{r_2}{r_1}}\left[(\theta_1 \log_e r_2 - \theta_2 \log_e r_1) - (\theta_1 - \theta_2)\log_e r\right]$$

$$= \frac{1}{2.3\log_{e10}\frac{12}{4}}\left[(427.7\times 2.3\times \log_{10} 12 - 20\times 2.3\times \log_{10} 4) - 407.7\times 2.3\log_{10} 8\right]$$

$$= \frac{1}{\log_{10} 3}\left[(427.7\log_{10} 12 - 20\log_{10} 4) - 407.7\log_{10} 8\right]$$

$$\theta = \frac{1}{0.4771}\left[(427.7\times 1.0792\times - 20\times 0.6021) - 407.7\times 0.9031\right]$$

$$= \frac{1}{0.4771}[461.4 - 12.04 - 368.1]$$

$$= \frac{81.29}{0.4771} = 170.3°C.$$

2. ***A thin metal pipe of 1 meter length and 1 cm radius carries steam at 100°C. This is covered by two layers of lagging. The thermal conductivity of outer layer, which is 2 cm thick, is 3.6 × 10⁻⁴ cal cm⁻¹ deg⁻¹ sec⁻¹, while that of the inner layer, which is 4 cm thick is 1.2 × 10⁻⁴ cal. cm⁻¹ deg⁻¹ sec⁻¹. If the outer surface of the lagging is at 30°C, find***

(a) ***the temperature of the cylindrical interface of the two lagging materials,***

(b) ***the mass of steam condensed per sec. (Given $\log_e 2 = 0.6231$).***

Solution : The heat flowing per sec. through the walls of a cylindrical tube is given by

$$Q = \frac{K2\pi l(\theta_1 - \theta_2)}{\log_e \frac{r_2}{r_1}}$$ [From eqn. (3) article 9.10]

Given that

For inner layer	For outer layer
$K_1 = 1.2 \times 10^{-4}$ C.G.S. units	
$K_1 = 3.6 \times 10^{-4}$ C.G.S. units	
$l = 100$ cm.	$l = 100$ cm.
$\theta_1 = 100°C$	$\theta_2 = 30°C$
$r_1 = 1$ cm.	$r_1 = 1 + 1 = 2$ cm.
$r_2 = 1 + 1 = 2$ cm.	$r_2 = 1 + 1 + 2 = 4$ cm.

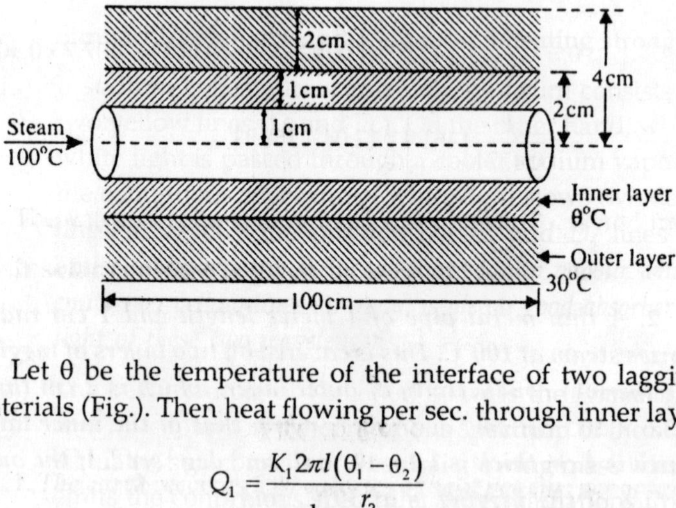

Let θ be the temperature of the interface of two lagging materials (Fig.). Then heat flowing per sec. through inner layer

$$Q_1 = \frac{K_1 2\pi l(\theta_1 - \theta_2)}{\log_e \frac{r_2}{r_1}}$$

$$= \frac{1.2 \times 10^{-4} \times 2 \times 3.14 \times 100 \times (100 - \theta)}{\log_e \frac{2}{1}} \quad \text{... (i)}$$

and the heat flowing per sec. through outer layer

$$Q_2 = \frac{K_2 2\pi l(\theta_1 - \theta_2)}{\log_e \frac{r_2}{r_1}}$$

$$= \frac{3.6 \times 10^{-4} \times 2 \times 3.14 \times 100 \times (\theta - 30)}{\log_e \frac{4}{2}} \quad \text{... (ii)}$$

But the heat flowing per sec. in the same through both layers of lagging, i.e.,

$$Q_1 = Q_2$$

$$\therefore \quad \frac{1.2\times10^{-4}\times2\times3.14\times100\times(100-\theta)}{\log_e 2}$$

$$= \frac{3.6\times10^{-4}\times2\times3.14\times100\times(\theta-30)}{\log_e 2}$$

or $\quad 1.2\times(100\text{-}\theta) = 3.6\cdot(\theta-30)$

or $\quad 100-\theta = 3(\theta-30))$

or $\quad \theta = 47.5°\text{C}.$

Now, let the mass of steam condensed per sec. be m gm. The heat taken away from the steam per sec = $m \times 540$ cal. = $Q_1 = Q_2$. Substituting in eqn. (i) the values of Q_1 and θ, we have

$$m\times540 = \frac{1.2\times10^{-4}\times2\times3.14\times100\times(100-47.5)}{\log_e 2}$$

$$m = \frac{1.2\times10^{-4}\times2\times3.14\times100\times52.5}{0.6931\times540}$$

$$=0.0106 \text{ gm}.$$

3. A ***rubber tube of length 20 cm, through which steam at 100°C is passing, is immersed in a calorimeter whose water equivalent is 15 gm and which contains 300 gm of water at 16°C. The temperature of water rises at the rate of 2°C per minute. If the outer and inner diameters of the tube are 1.0 cm and 0.6 cm respectively, calculate the thermal conductivity of rubber.***

Solution : The radial rate of heat flow through a tube of length l and internal and external radii r_1 and r_2 is

$$\frac{dQ}{dt} = \frac{2\pi l(\theta_1-\theta_2)}{\log_e\left(\frac{r_2}{r_1}\right)} = \frac{2\pi Kl(\theta_1-\theta_2)}{2.3026\log_{10}\left(\frac{r_2}{r_1}\right)}$$

where θ_1 and θ_2 are the inside and outside temperatures of the tube. If it raises the temperature of m gm water contained in a calorimeter of water equivalent W by $\Delta\theta$, then

$$\frac{dQ}{dt} = (m+W)\Delta\theta$$

Equating the above two expressions

$$\frac{2\pi Kl(\theta_1 - \theta_2)}{2.3026\log_{10}\left(\frac{r_2}{r_1}\right)} = (m+W)\Delta\theta$$

or
$$K = \frac{(m+W)\Delta\theta 2.3026\log_{10}\left(\frac{r_2}{r_1}\right)}{2\pi Kl(\theta_1 - \theta_2)}$$

Given $m + W = 300 + 15 = 315$ gm.

$$\Delta\theta = 2°\text{C/minute} = \frac{1}{30}\,°\text{C/sec.}$$

$$r_2 = 1 \text{ cm}, r_1 = 0.6 \text{ cm}, l = 20 \text{ cm},$$

$$\theta_1 = 100°\text{C}, \theta_2 = 16°\text{C}$$

$$\therefore \quad K = \frac{315\left(\frac{1}{30}\right) \times 2.3026 \times \log_{10}\left(\frac{1}{0.6}\right)}{2 \times 3.14 \times 2 \times (100 - 16)}$$

$$= \frac{10.5 \times 2.3026 \times 0.2217}{2 \times 3.14 \times 20 \times 20 \times 84}$$

$$= 5.1 \times 10^{-4} \text{ cal cm}^{-1}\text{ sec}^{-1}\,°\text{C}^{-1}.$$

Temperature Distribution in a Spherical Shell Heated at the Centre (Radial Flow of Heat) : Consider a spherical shell of internal and external radii r_1 and r_2 respectively.

Let a source of heat, e.g., an electric heating element, be placed at the centre O of the shell. The heat is conducted radially through the shell and lost subsequently to the surroundings from the other surface. Let θ_1 and θ_2 be the constant temperatures of the inside and outside of the shell in steady state (Fig.).

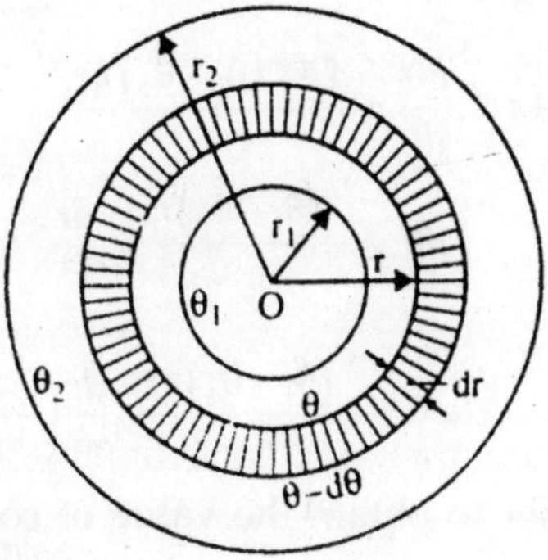

Consider an elementary shell of radius r and thickness dr having temperature θ and $(\theta - d\theta)$ on its inner and the outer surfaces respectively.

The amount of heat flowing per sec. through this shell

$$Q = -KA\frac{d\theta}{dr} = -K\,4\pi r^2 . \frac{d\theta}{dr} \quad ...(i)(\because A = 4\pi r^2)$$

$$\frac{dr}{r^2} = -\frac{4\pi K}{Q}.d\theta.$$

Integrating both sides between their respective limits and remembering that Q is independent of r, we get

$$\int_{r_1}^{r_2} \frac{dr}{r^2} = -\frac{4\pi K}{Q}\int_{\theta_1}^{\theta_2} d\theta$$

or
$$\frac{1}{r_1} - \frac{1}{r_2} = \frac{4\pi K(\theta_1 - \theta_2)}{Q}$$

or
$$K = \frac{Q(r_1 - r_2)}{4\pi r_1\, r_2(\theta_1 - \theta_2)} \quad ... (ii)$$

From it the value of conductivity K can be calculated. It may be employed to find the conductivity of substances which are available in powder form, such as sand, clay, charcoal, etc.

Now the temperature distribution in the spherical shell may be obtained. Equating the values of Q from equations (i) and (ii), we get

$$-K.4\pi r^2.\frac{d\theta}{dt}=\frac{4\pi K(\theta_1-\theta_2)r_1r_2}{r_2-r_1}$$

$$d\theta=-\frac{(\theta_1-\theta_2)r_1r_2}{r_2-r_1}.\frac{dr}{r^2}$$

Integrating, $$\theta=\frac{(\theta_1-\theta_2)r_1r_2}{r_2-r_1}.\frac{dr}{r^2}\left(-\frac{1}{r}\right)+C \quad \text{... (iii)}$$

Now in order to obtain the value of constant C, we have at

$$r=r_1,\quad \theta=\theta_1,$$

$$\therefore \quad \theta_1=\frac{(\theta_1-\theta_2)r_1r_2}{r_2-r_1}.\frac{1}{r_1}+C.$$

or $$C=\theta_1-(\theta_1-\theta_2)\frac{r_2}{r_2-r_1}$$

Substituting it in equation (iii), we get

$$\theta=\frac{(\theta_1-\theta_2)r_1r_2}{r(r_2-r_1)}+\theta_1-(\theta_1-\theta_2)\frac{r_2}{r_2-r_1}$$

$$\theta=\frac{r_2}{r_2-r_1}\left\{\frac{r_1r_2(\theta_1-\theta_2)}{r}+r_2\theta_1-r_1\theta_1+r_2\theta_2-r_2\theta_1\right\}$$

or $$=\frac{1}{r_2-r_1}\left\{\frac{r_1r_2(\theta_1-\theta_2)}{r}+r_2\theta_2-r_1\theta_1\right\}$$

This gives the temperature of spherical surface of radius r.

PROBLEMS

1. ***Two thin concentric spherical shells of radii 5 cm and 10 cm respectively have their annular cavity filled with charcoal powder. When energy is supplied at the rate of 10.5 watt to a heater at the centre, a temperature difference of 60°C is set***

up between the shells. Find the thermal conductivity of charcoal. (J = 4.2 joule /cal).

Solution : The thermal conductivity of charcoal will be given by

$$K = \frac{Q(r_1 - r_2)}{4\pi r_1 r_2 (\theta_1 - \theta_2)}$$

Here, heat supplied per sec. Q = 10.5 × Joule/sec.

$$= \frac{10.5}{J} \text{cal/sec.} \quad = \frac{10.5}{J} \text{cal/sec.}$$

Also given r_2 = 10 cm, r_1 = 5 cm, $(\theta_1 - \theta_2)$ = 60°C

$$K = \frac{\frac{10.5}{4.2}(10-5)}{4 \times 3.14 \times 10 \times 5 \times 60}$$

$$= 3.3 \times 10^{-4} \text{ cal. cm}^{-1} \text{ deg}^{-1} \text{ sec}^{-1}.$$

Propagation of Heat Wave in an Insulated Rod with End Heated Sinusoidally (Periodic Flow of Heat) : Uptill now we have confined our discussions to the steady flow of heat along a rod. Now we shall consider the case where one end of metallic bar is alternately heated and cooled, i.e., the supply of heat from the source is not steady, but periodic. Then the temperature of different points in the bar will also vary periodically. Whenever the end is heated a heat wave flows outward from it, but when it is cooled a heat wave travels towards it and thus two heat waves flow alternately along the rod in opposite directions. If, however, this alternate heating and cooling is continued for a sufficiently long time the temperature variation at any point will acquire a fixed character so that the mean temperature at every point remains steady, called the steady periodic state.

Confining ourselves to an insulated well-lagged rod, so that there is no loss of heat due to radiations, the problem reduces to the propagation of periodic heat only in one direction. Let the periodic variation of temperature θ be represented by $\theta = \theta_0 \cos wt$, θ_0 being the amplitude of variation and ω the angular frequency of the heat or temperature wave. The fluctuations of temperature along the bar can be represented by a sine curve of damped type shown in the Fig., where the amplitude of temperature variation goes on decreasing as the distance from the hot end increases.

Let the direction of heat flow be along the X-axis, then as derived earlier [article 9.8 eq. (v)], the Fourier equation representing unidirectional heat flow along a lagged bar is,

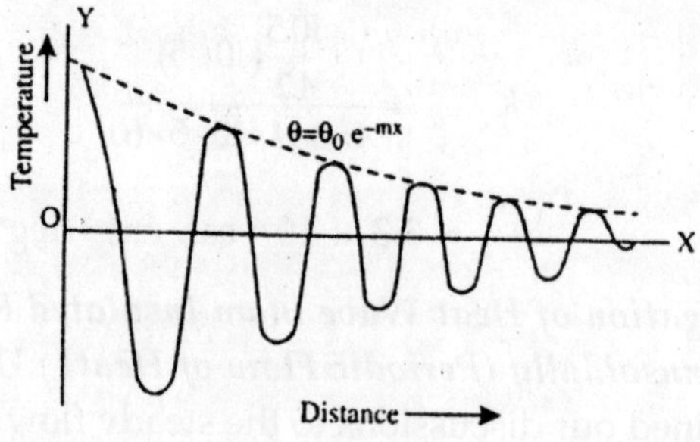

$$\frac{d\theta}{dt} = h\frac{d^2\theta}{dx^2}, \quad \text{... (1)}$$

where $h = \frac{K}{\rho}$ is the diffusivity of the bar.

Also, we have $\theta = \theta_0 \cos \omega t$, at $x = 0$

and since there is no increase in temperature as we get further into the solid, we must have

$$\theta \neq \infty \quad \text{at} \quad x = \infty$$

With these boundary conditions let us solve the differential eq. (1). Since temperature variation is a simple harmonic function of time, we can assume a solution

$$\theta = Ae^{\alpha x + i\beta t} \quad \text{... (2)}$$

where $i = \sqrt{-1}$ and A, α and β are constants.

Substituting in eqn. (1) the value of $\frac{d\theta}{dt} = Ai\beta e^{\alpha x+i\beta t}$

and $$\frac{d^2\theta}{dx^2} = A\alpha^2\beta e^{\alpha x+i\beta t}$$

so that from (1), we get

$$Ai\beta e^{\alpha x+i\beta t} = hA\alpha^2 e^{\alpha x+i\beta t}$$

or $$i\beta = h\alpha^2$$

so that $$\alpha = \pm\sqrt{\left(\frac{i\beta}{h}\right)}$$

and since $(1 + i)^2 = 2i$, we have $\sqrt{i} = \frac{1+i}{\sqrt{2}}$

$$\therefore \qquad \alpha = \pm\sqrt{\left(\frac{i\beta}{2h}\right)}(1+i)$$

and hence $$\theta = Ae^{\pm\sqrt{(\beta/2h)}(1+i)x+i\beta t}$$

Since $\theta \neq \infty$, when $x \neq \infty$, we reject the +ve sign, before the square root, and hence

$$\theta = Ae^{-\sqrt{(\beta/2h)}(1+i)x+i\beta t}$$

$$= Ae^{-\sqrt{(\beta/2h)}.x+i(\beta t-\sqrt{(\beta/2h)}.x)}$$

It may also be put in the form

$$\theta = Ae^{-\sqrt{(\beta/2h)}.x}\left[\cos\left\{\beta t - \sqrt{\left(\frac{\beta}{2h}\right\}}.x + i\sin\left\{\beta t - \sqrt{\left(\frac{\beta}{2h}\right)}.x\right]$$

The imaginary terms in the expression are of no interest to us and hence may be neglected. Therefore,

$$\theta = Ae^{-\sqrt{(\beta/2h)}.x}\cos\beta t - \left\{\beta t - \sqrt{\left(\frac{\beta}{2h}\right)}.x\right\} \quad \text{... (3)}$$

At $x = 0$, it becomes $\theta = A \cos \beta t$.

But we have assumed, at $x = 0$, $\theta = \theta_0 \cos \omega t$.

Consequently, $A = \theta_0$ and constant β must be identical with ω of the periodic motion applied.

Substituting the values of A and β in expression (3) for θ, we have

$$\theta = \theta_0 e^{-\sqrt{(\omega/2h)}.x} \cos\left[\omega t - \sqrt{\left(\frac{\omega}{2h}\right)} x\right] \quad \text{... (4)}$$

In order to give it a more general form, we introduce an arbitrary constant ε in the cosine term and hence

$$\theta = \theta_0 e^{-\sqrt{(\omega/2h)}x} \cos\left[\omega t - \sqrt{\left(\frac{\omega}{2h}\right)} x - \varepsilon\right] \quad \text{... (5)}$$

It represents a progressive wave, travelling with a velocity $\sqrt{(2\omega h)} = \sqrt{\left(\frac{4\pi h}{T}\right)}$, and of amplitude of fluctuations

$$\theta_0 e^{-\sqrt{(2\omega h}\, x} = \theta_0 e^{-mx},$$

where $m = \sqrt{\left(\frac{\omega}{2h}\right)}$, is the damping factor.

Substituting the values of w and h, the velocity of heat wave

$$v = \sqrt{(2\omega h)} = \sqrt{2 \times \frac{2\pi}{T} \times \frac{k}{\rho c}} = \left(\frac{4\pi k}{T\rho c}\right)$$

or

$$\frac{\lambda}{T} = \sqrt{\frac{4\pi k}{T\rho c}}$$

or

$$K = \frac{\lambda^2 \rho c}{4\pi T} \quad \text{... (6)}$$

Angstrom's Method to Determine the Thermal Conductivity of a Metal Bar : Angstrom applied the method of periodic heat flow to determine the thermal conductivity of a metal bar. He used a long bar and enclosed one end of it in a chamber through which steam and cold water at known temperatures could be alternately passed. The bar was heated for 12 minutes and then cooled for the same time. Thus the periodic time for heat flow T = 24 minutes. It was repeated a number of times so that ultimately the mean temperature at any point of the bar becomes constant. Then the temperatures at two points on the bar were observed by means of thermocouples. The distance between these two points gives the wavelength l of the heat wave.

Knowing λ and T, the value of K for the bar can be determined from eq. (6) of section 9.12, the values of ρ and c for metal being known from standard tables.

Propagation of Temperature Waves in the Earth's Crust : As the earth rotates about its axis, its surface is heated up during day time and cooled down during the night. It is thus subjected to a periodic heating and cooling and as a result a temperature wave with a period of twenty-four hours is propagated into the interior of the earth. This wave is called the dirunal wave.

Besides this the earth rotates round the sun, any point on its surface receives a larger amount of heat in the summer than in the winter. It gives rise to a second wave, of period one year, travelling into the interior of the earth. This wave is known as the annual wave.

Assuming that these waves are simple harmonic to a first approximation; the phenomenon may be compared to that of a lagged bar periodically heated and cooled at one end and the Fourier differential equation

$$\frac{d\theta}{dt} = h\frac{d^2\theta}{dx^2}$$

may be applied to this case. Its solution as derived in eqn. (5) of section 9.12 is

$$\theta = \theta_0 e^{-\sqrt{(\omega/2h)}x} \cos\left[\omega t - \sqrt{\left(\frac{\omega}{2h}\right)}x - \varepsilon\right] \quad \text{... (1)}$$

where ω, known as pulsatance, represents the number of vibrations in 2π units of time, i.e., $\omega = \frac{2\pi}{T}$, T being the periodic time.

Eqn. (1) represents a progressive wave. We also know that a progressive harmonic wave of wavelength λ may be expressed as

$$y = A\cos 2\pi\left(\frac{t}{T} - \frac{x}{\lambda}\right). \quad \text{... (2)}$$

In wave motion, we also introduce a quantity k, known as undulatance, which expresses the number of waves in $2a$ units of length, $k = \frac{2\pi}{\lambda}$.

Substituting $\frac{2\pi}{T} = \omega$ and $\frac{2\pi}{\lambda} k$ in eqn. (2), it becomes

$$y = A \cos (\omega t - kx).$$

The velocity of propagation of the wave

$$v = \frac{\lambda}{T} = \frac{2\pi}{k}\Big/\frac{2\pi}{\omega} = \frac{\omega}{k} = \frac{\text{Pulstance}}{\text{Undulatance}}.$$

Therefore, velocity of propagation of heat wave, from eqn. (1)

$$v = \frac{\omega}{\sqrt{\left(\frac{\omega}{2h}\right)}} = \sqrt{(2\omega h)}. \quad \text{... (4)}$$

Substituting $v = \frac{\lambda}{T}$ and also the values $\omega = \frac{2\pi}{T}$ and $h = \frac{K}{\rho c}$, it becomes

$$\frac{\lambda}{T} = \sqrt{\left(2.\frac{2\pi}{T}.\frac{K}{\rho c}\right)}$$

or
$$\frac{\lambda^2}{T^2} = \frac{4\pi K}{T\rho c}$$

$$\therefore \quad K = \frac{\lambda^2}{4\pi T}\rho c \qquad \text{... (5)}$$

The periodic variations of temperature in the earth can be found by embedding thermometers in the earth at various depths and thus knowing the wavelength λ, the conductivity K of earth crust can be calculated from relation (5) after substituting for time period T, density ρ and specific heat c.

PROBLEMS

1. *The annual range of fluctuation of temperature in the earth at depths 6 and 24 feet are 5.6°C and 0.7°C respectively. Calculate the velocity of propagation of temperature wave in the earth. Also find the diffusivity of the earth's crust.*

Solution : The equation for a temperature wave, derived in article 9.12 is

$$\theta = \theta_0 e^{-\sqrt{(\omega/2h)}.x} \cos\left[\omega t - \sqrt{\left(\frac{\omega}{2h}\right)}x - \varepsilon\right].$$

The amplitude of temperature fluctuation at any depth x is

$$\theta = \theta_0 e^{-\sqrt{(\omega/2h)}x} = \theta = \theta_0 e^{-mx} \quad m = \sqrt{\left(\frac{\omega}{2h}\right)}$$

Given θ = 5.6°C at x = 6 feet and θ = 0.7°C at x = 24 feet.

$$\therefore \quad 5.6 = \theta_0\ e^{-6m} \text{ and } 0.7 = \theta_0\ e^{-24m}$$

Dividing the first equation by the second

$$\frac{5.6}{0.7} = \frac{\theta_0 e^{-6m}}{\theta_0 e^{-24m}} \text{ or } 8 = e^{18m}$$

$$\therefore \quad \log_e 8 = 18\text{ m or } 18\text{ m} = 2.3\log_{10} 8 = 2.3 \times 0.931.$$

Hence 18 m = 2.077 or m = 0.1154

Now the velocity of propagation of heat wave (eqn. 4, article 9.14) is given by

$$v=(2\omega h)=\omega\sqrt{\left(\frac{2h}{\omega}\right)}=\frac{\omega}{m}.$$

But $\omega=\frac{2\pi}{T}$ and here T = 1 year = 365 days.

$$\therefore \qquad \omega=\frac{2\pi}{365}\text{ radians/day}$$

$$v=\frac{\omega}{m}=\frac{2\pi}{365\times 0.1154}=0.15\text{ feet per day.}$$

Now diffusivity

$$h=\frac{\omega}{2m^2} \qquad \left[\because \quad m=\sqrt{\frac{\omega}{2h}}\right]$$

$$=\frac{2\pi}{365\times(0.1154)^2}=0.654\text{ feet per day.}$$

2. *In a periodic flow of heat along an iron bar, the periodic time is 4 minute. If the temperature maximum travels 6 cm in one minute, calculate the thermal conductivity of iron. Density of iron = 7.8 gm /cm³, specific heat of iron = 0.11 cal /gm °C.*

Solution : The velocity of heat wave along the iron bar

$$v=(2\omega h)=\sqrt{2\times\frac{2\pi}{T}\times\frac{k}{\rho c}}$$

$$\therefore \qquad v^2=\frac{4\pi k}{T\rho c}$$

Here $\qquad v$ = 6 cm/minute = 0.1 cm/sec.

T = 4 × 60 = 240 sec.

c = 0.11 cal/gm°C,

ρ = 7.8 gm./cm³

Therefore, thermal conductivity of iron

$$K=\frac{v^2T\rho c}{4\pi}$$

$$= \frac{(0.1)^2 \times 240 \times 7.8 \times 0.11}{4 \times 3.14}$$

$$= 0.1639 \text{ cal cm}^{-1} \text{ sec}^{-1} \text{ °C}^{-1}.$$

Experimental Determination of Thermal Conductivity : We will now deal with the experimental determination of conductivity of both good and bad conductors.

(A) Methods for Good Conductors

Searle's Method : It is the simplest method for the determination of thermal conductivity of the good conductors like metals. Figure below depicts the experimental arrangement of the apparatus in which *XY* is the cylindrical bar of the metal whose coefficient of thermal conductivity is to be determined. Its end *X* is enclosed in a steam jacket through which steam is passed continuously. Near the end *Y*, a copper tube is wound helically through which cold water is allowed to pass. The bar is well covered with felt or some other non-conducting material to avoid loss of heat from it.

When steam is passed through the jacket for quite some time, heat is conducted along the bar and a steady state is reached. Then the thermometers $\theta_1, \theta_2, \theta_3, \theta_4$ will record constant temperatures. The amount of heat passing from section *P* to *Q* then becomes constant and the same amount of heat flows through the section at *Y*. The mass of water flowing out of the copper tube in a known interval of time is then noted.

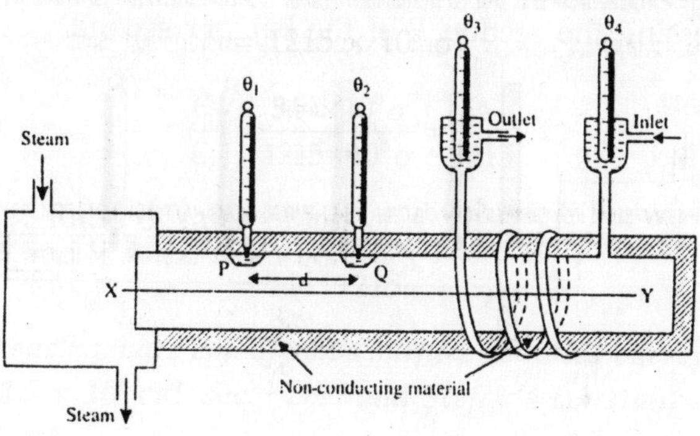

If m gm. is the mass of water collected in a time t sec., then amount of heat gained by the water in this time

$$= m\,(\theta_4 - \theta_3) \qquad \text{... (i)}$$

If A is the area of cross section of the bar, the amount of heat passing through the section PQ of the bar in time t

$$= \frac{KA(\theta_1 - \theta_2)t}{d} \qquad \text{...(ii)}$$

Equating (i) and (ii)

$$\frac{KA(\theta_1 - \theta_2) \times t}{d} = m(\theta_4 - \theta_3)$$

$$\therefore \qquad K = \frac{md(\theta_4 - \theta_3)}{At(\theta_1 - \theta_2)}$$

All other quantities being known, the coefficient of thermal conductivity K of the material of the bar can be determined.

Forbe's Method: Forbe's method is used to determine the absolute conductivity of good conductors directly in terms of the definition of K given by the fundamental relation,

$$Q = KA\left(\frac{d\theta}{dx}\right)t.$$

Theory : Consider a long rectangular metal bar AB of uniform area of cross-section heated at one end (Fig.). The bar is not lagged here to prevent radiation but is exposed to atmosphere so that heat escapes from its surface throughout. When steady state is reached, after some time, the amount of heat crossing the section P of the bar per second

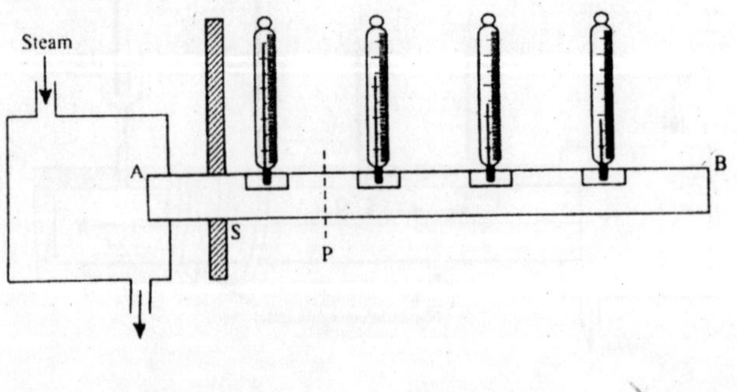

$$Q_P = KA\left(\frac{d\theta}{dx}\right)_P$$

will be equal to the amount of heat lost by radiation by the portion of the bar between section *P* and the end *B* which is

$$= \int_P^B (\rho\ Adx)\ c\left(\frac{d\theta}{dt}\right)$$

Thus $$KA\left(\frac{d\theta}{dx}\right)_P = \int_P^B (\rho\ Adx)\ c\left(\frac{d\theta}{dt}\right)$$

or $$K = \frac{\rho c \int_P^B \left(\frac{d\theta}{dt}\right) dx}{\left(\frac{d\theta}{dx}\right)_P} \qquad \text{... (i)}$$

where ρ and c are respectively the density and specific heat of the material of the bar, $\frac{d\theta}{dt}$ is the rate of fall of temperature and $\left(\frac{d\theta}{dt}\right)_P$ is the temperature gradient at *P*.

To find the value of $\left(\frac{d\theta}{dx}\right)_P$ and $\left(\frac{d\theta}{dt}\right)dx$, the experiment is divided into two parts :

Statical Part : The long metal bar shown in the fig., is heated at one end and the temperatures recorded by various equidistant thermometers in steady state are read. A graph is plotted between the steady temperatures 9 and the distance × from the hot end (Fig.). A tangent is drawn corresponding to the point P on the , exponential curve so obtained. The slope of the curve gives the value of $\left(\frac{d\theta}{dx}\right)_P$. Thus

$$\left(\frac{d\theta}{dx}\right)_P = \tan\alpha$$

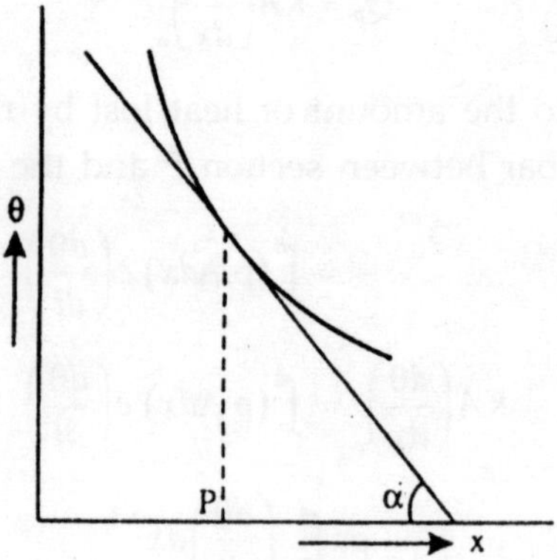

Dynamical Part : The original bar, or a small bar identical with the previous one in cross-sectional area and in the nature of its surface, is heated to the steam temperature. It is then allowed to cool in exactly the same surroundings as obtained during the statical part by allowing it to rest on two non-conducting knife-edges (Fig.). The temperature of the bar is noted at regular intervals of time by a thermometer placed in a hole at its centre. A graph is plotted between the temperature θ and the time t (Fig.). $\frac{d\theta}{dt}$ is determined for various values of θ by drawing tangents at a number of points on the curve so obtained. The slopes of these tangents gives the value of $\frac{d\theta}{dt}$ at different temperatures θ. For these temperatures θ, the corresponding values of × are determined from the graph between θ and x (Fig.). We thus get a set of values of $\frac{d\theta}{dt}$ for different values of x.

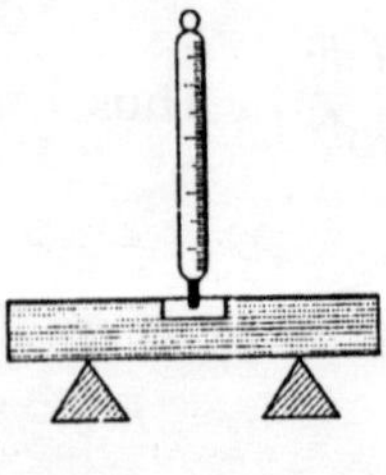

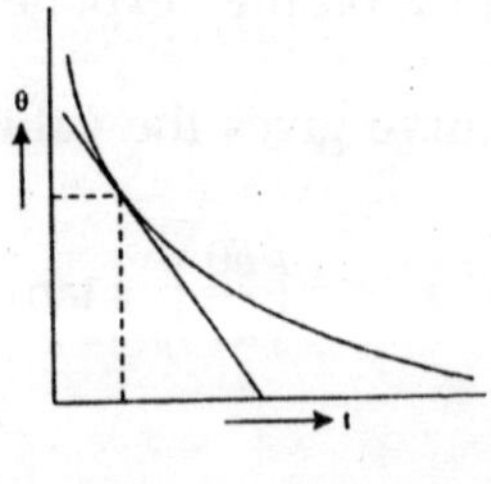

A graph is now plotted between $\frac{d\theta}{dt}$ and x (Fig.). An ordinate is erected corresponding to that at section P of the bar and the curve is produced to meet the axis of x which it will naturally do corresponding to the end point B. Then the shaded area, bounded between the curve, the x-axis and the ordinate at P, will give the value of the integral $\int_P^B \left(\frac{d\theta}{dt}\right) dx$, which can be measured directly from the graph or with the help of a planimeter.

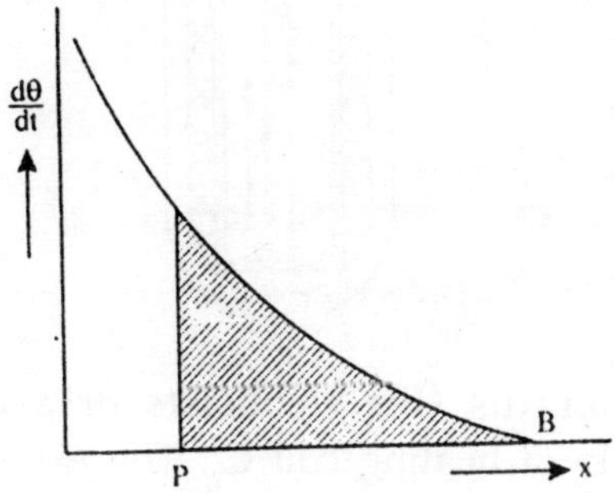

Thus obtaining the values of $\left(\frac{d\theta}{dx}\right)_P$ tan α and $\int_P^B \left(\frac{d\theta}{dt}\right) dx$ from the statical and dynamical parts of the experiment, the value of K can be calculated from eq. (1).

Drawbacks of Forbe's Method

The method, however, suffers from the following main drawbacks

(1) The method is tedious and takes a long time.

(2) The specific heat c of the bar does not remain constant at different temperatures as per assumption.

(3) The distribution of temperatures along the bar in the statical and dynamical parts are not the same.

(4) There is hardly any justification for assuming Newton's law of cooling to be true for different temperature excesses.

(5) The whole length of the bar does not acquire a true steady state.

Lee's Method : Lee's method is useful for measuring the conductivity of metals at low temperatures. It is also suitable for investigating the variation of thermal conductivity with temperature.

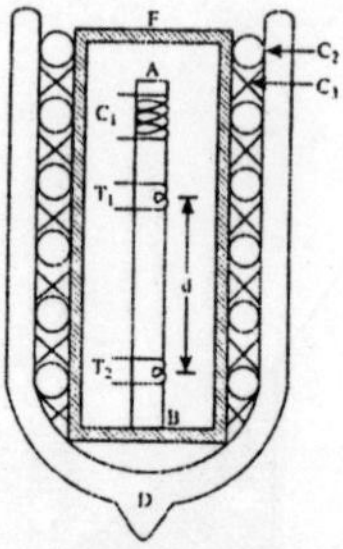

The apparatus (Fig.) consists of a copper frame *F* surrounded by a heating coil C_2. The experimental bar *AB*, about 7 cm. long and 0.5 cm in diameter, is heated vertically in a socket provided at the bottom of the frame F. The upper end *A* of the bar is heated by a heating coil C_1. T_1 and T_2 are two platinum resistance thermometers which can be made to slide up and down the bar and adjusted at any desired distance *d* apart. The whole arrangement is placed inside a Dewar flask *D*, in which any desired temperature can be produced and maintained.

In order to determine the conductivity *K* of the experimental bar, it is first cooled to some degrees below that temperature at which the thermal conductivity is to be measured. It is done by first pouring liquid air into the flask and then removing the remaining after some time The evaporation of the liquid air leaves apparatus cool and dry, Current is now started in coil C_2 and switched off when the whole apparatus is brought to the desired temperature. Now the experimental bar is heated up by passing a known current through the heating coil C_1.

This current is continued till the steady state is reached after a time t. Then the constant temperatures θ_1 and θ_2 shown by T_1 and T_2 are noted. If E (volts) is the potential difference across the heating coil C_1 and I (Ampere) is the current flowing through it, then

Heat produced by the current = $E.I.t$ Joule

$$= \frac{E.I.t}{4.2} \text{ calories.}$$

In steady state, it should be equal to the quantity of heat flowing across the length d of the bar. Hence

$$\frac{E.I.t}{4.2} = \frac{KA(\theta_1 - \theta_2)t}{d}$$

or

$$K = \frac{E.I.d}{4.2A(\theta_1 - \theta_2)} \quad \text{... (i)}$$

where A is the area of cross section of the bar.

Source of Error and Correction : There is an important source of error in this experiment. When the end A of bar AB is heated by coil C_1, the copper frame carrying it is also warmed up which then tends to raise the temperature of the bar itself by radiating heat to it. This backward flow of heat, keeps on warming the bar and thus the steady state is never truly reached. Due to it, observed $(\theta_1 - \theta_2)$ will be less than the its actual value.

To correct this error, another heating coil C_3 of resistance equal to that of C_1 is wound round frame F. Before performing the main experiment the same current I is first passed through coil C_3 (without passing through C_1 and the temperature difference $d0_1$ shown by T_1 and T_2 is noted in steady state. The same procedure is repeated after the main experiment is over and again the temperature difference $d\theta_2$ shown by T_1 and T_2 *is* noted. Then

$$\text{Correction to be applied} = \frac{d\theta_1 - d\theta_2}{2}$$

Hence the correct temperature difference between T_1 and T_2

$$= (\theta_1 - \theta_2) + \left(\frac{d\theta_1 - d\theta_2}{2}\right)$$

This, on substituting in place of $(\theta_1 - \theta_2)$; eq. (1) gives the thermal conductivity

$$K = \frac{E.I.d}{4.2A\left[(\theta_1 - \theta_2) + \left(\frac{d\theta_1 - d\theta_2}{2}\right)\right]} \quad \text{... (ii)}$$

By this method, Lee found K for different metals between –170°C and 18°C and showed that K increases with rise in temperature between –170°C and 100°C for silver, copper, zinc and aluminium, whereas it decreases for iron and lead.

Method for Poor Conductors (Lee's Disc Method) : This method is used to determine the thermal conductivity of poor conductors like ebonite, glass, cork, wood, etc. The method is illustrated in the figure. Here the specimen is taken in the form of a thin disc *D* about 4 cm in diameter and 2 mm in thickness and is sandwiched in between two similar copper plates C_2 and C_3. A thin layer of glycerine is applied to the two faces of *D* to secure good thermal contacts with C2 and C_3. An insulated flat heating coil *H* of platinum is placed between C_2 and another thinner (~1 mm thick) copper disc C_1. Thermocopules T_1, T_2 and T_3 are inserted in tiny holes drilled in the edges of the three copper plates to measure temperature. The whole assembly is suspended in a constant temperature enclosure by non-conducting threads.

Now, the current in the heating coil *H is* started and in the steady state the temperature θ_1, θ_2 and θ3 of the three copper discs are read. The values of the heating current I and its voltage V are also noted on the ammeter A and voltmeter V connected to the heating coil H. Then quantity of heat passing through the specimen disc D in time t sec.

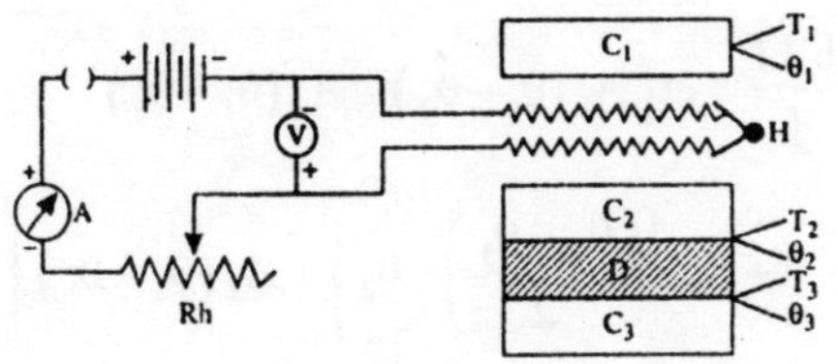

$$= KA.\frac{\theta_2 - \theta_3}{d}.t \quad \text{... (i)}$$

where A is the area of cross section and d the thickness of the disc D.

In the steady state, it must obviously be equal to the quantity of heat which has entered in the same time t from the exposed surfaces of copper disc C_3 and that of D itself. Since the quantity of heat crossing any plane of D is the average of that entering it and leaving it, we take into calculation only half the heat emitted by it. If, therefore, e be the surface emissivity of polished surface of the discs, A_D the area of exposed surface of specimen disc D and A_3 the exposed surface area of disc C_3, then quantity of heat emitted in time t sec from the exposed surfaces of D and C_3

$$= e.t\left[\frac{A_D}{2}\left(\frac{\theta_2 - d_3}{d} - \theta_0\right) + A_3(\theta_3 - \theta_D)\right] \quad \text{... (ii)}$$

where θ_0 is the temperature of the surrounding enclosure and $\frac{\theta_2 - \theta_3}{2}$ is the average temperature of D.

Equating equations (i) and (ii), we have

$$= KA.\frac{\theta_2 - \theta_3}{d}.t = e.t\left[\frac{A_D}{2}\left(\frac{\theta_2 - d_3}{d} - \theta_0\right) + A_3(\theta_3 - \theta_D)\right] \quad \text{... (iii)}$$

To evaluate *e*, use is made of the fact that in the steady state, all the heat that is produced by a current of *I* ampere at *E* volts in time *t is* lost by radiation from the exposed surface of the whole assembly. Hence

$$\frac{E.I.t}{J} = et\left[A_1(\theta_1 - \theta_2) + A_2(\theta_1 - \theta_0)\right.$$

$$\left. + A_D\left\{\left(\frac{\theta_2 - \theta_3}{2}\right) - \theta_0\right\} + A_3(\theta_3 - \theta_0)\right] \quad \text{... (iv)}$$

where $J = 4.2$ Joule per cal.; A_1, A_2 and A_3 being the area of the exposed surfaces of C1, C_2 and C_3 respectively.

Using the value of *e* from (iv), the value of thermal conductivity K of the specimen can be calculated.

This method cannot be used for good conductors because in case of a good conductor, the difference of temperature between the two faces will be too small to be measured.

5

Radiant Heat

We receive heat from the sun which is millions of kilometers away from the earth. After a few thousand kilometers, there is only empty space with no material medium. If at all there is, the density does not exceed one or two molecules per cubic centimeter. The presence of matter of this density can be entirely ignored. Therefore, the process, by which we receive heat from the sun, cannot be either conduction or convection which require an intervening material medium. The process is termed as radiation. Heat travelling by radiation is called radiant heat or thermal radiation.

Thermal Radiations

During this process, heat from the sun passes through air around the earth; but the air is not heated. Maxwell defined *radiation as the transfer of heat from a hot body to a cooler body without appreciable heating of the intervening medium (or space).*

This statement does not mean that an intervening medium will not be heated when thermal radiation falls on it or passes through it. In fact, the bodies may absorb radiant heat to a greater or lesser degree when the radiations fall on them or pass through them and thereby get heated to a greater or lesser extent. The examples of such substances are glass, wood, metal etc. On the other hand, some substances allow radiant heat to pass through them without being heated to any great extent. The examples of such substances are quartz, rock salt, fluorspar etc.

Conclusively, we can say that when heat is propagated by radiation, no material medium is necessary for the transmission and if there is any material medium, it is not necessary that it should first get itself heated (as in the case of conduction and convection) before it could assist the propagation of thermal radiation. Moreover, it had been shown that heat is a form of energy like light. The propagation of heat by radiation consists merely in transference of energy which can take place even in empty space. The investigations show that thermal radiation has following properties :

(1) Thermal radiation travels through empty space with the velocity of light.

(2) Thermal radiation, like light, travels in straight lines.

(3) Thermal radiation, like light, obeys the law of inverse square.

(4) Thermal radiation is reflected and refracted like light.

(5) Thermal radiation exhibits the phenomenon of interference, diffraction and polarisation like light.

From the above properties of thermal radiation, it is clear that it has the same nature as light. The only difference is that average wavelength of thermal radiation is greater than that of visible light. Therefore, thermal radiation is called the infra-red radiation. The spectrum of radiations given out by hot body may be formed, by using the prisms of rock–salt, KC1 etc. or diffraction grating of large spacings, just as the spectrum of visible radiation is formed by using ordinary glass prism or grating.

The thermal radiations cannot be detected by the eye or the photographic plate but they can be detected by the bolometer, which depends on the change of electrical resistance with temperature or thermopile which is just a series of thermocouples connected properly to give an e.m.f. proportional to incident energy.

Some Definitions

Total Energy Density : *Total energy density of radiations at any point is the total radiant energy per unit volume around that point for all the wavelengths taken together.* This is generally expressed by u, its unit is joule m^{-3} or erg. cm^{-3}.

Spectral Energy Density : *Spectral energy density for a particular wavelength is the energy per unit volume per unit range of wavelength.* This is denoted by u_λ.

Total Emissive Power or Emissivity : *The total emissive power of a body is the radiant energy emitted per unit time per unit surface area of body for all wavelengths taken together.* It is denoted by E.

Spectral Emissive Power: *The spectral emissive power of a body at a particular wavelength is the radiant energy emitted per unit time per unit surface area of the body within a unit wavelength range.* It is denoted by E_λ.

Absorptive Power or Absorptivity : *The absorptive power of a body at a particular temperature and for a particular wavelength is defined as the ratio of the radiant energy absorbed per unit surface area per unit time to the total energy incident on the same area of the body in unit time within a unit wavelength range.* It is denoted by a_λ.

From the above definitions it is obvious that

$$u = \int_0^\infty u\lambda \, d\lambda \quad \text{... (i)}$$

$$E = \int_0^\infty E\lambda \, d\lambda. \quad \text{... (ii)}$$

Relative Emittance **:** The ratio of emittance of a surface to the emittance of a black body is called relative emittance.

If E' is the emittance of a surface and E the emittance of a black body, then relative emittance of the surface

$$e = \frac{E'}{E}$$

Prevost's Theory of Exchanges

When we stand near a hot body, we experience the sensation of heat which can be explained by the fact that the hot body radiates heat; but if we stand near a block of ice, we experience the sensation of cold which was explained on the assumption that ice emits cold radiations. To get over this difficulty, Prevost pointed out that the conception, the ice emits cold radiations, was incorrect and proved that *thermal radiation is essentially an exchange process.*

According to Prevost, every body emits radiations continuously to its surroundings at a rate which depends only on the nature of its surface and temperature and at the same time it receives radiation from its surroundings at a rate depending on the surface and the temperature of surroundings.

A body at a temperature higher than the surroundings radiates more quantity of heat per unit time to the surroundings than it receives from the surroundings. Consequently, the temperature of the body will fall. The exchange of energy continues even after the body emits as much energy per unit time as it receives and hence the temperature of the body remains constant. When the temperature of the body is equal to that of the surroundings, we say that the body is in thermal equilibrium with the surroundings. Similarly, when a body is at a temperature lower than that of surroundings it will radiate less heat and absorb more heat.

Thus, all bodies at all temperatures are in a state of dynamical thermal equilibrium when they are at the same temperature. The quantity of heat radiated by a body decreases with the fall in temperature. When the body is at absolute zero, it will not radiate energy.

Radiation in an Isothermal (Uniform Temperature) Enclosure : In general, the nature of radiation and the amount of radiation per unit area per second i.e. the quality and quantity of radiation depends on the temperature as well as on the nature of the surface. If, however, *we consider a non-conducting enclosure maintained at a constant temperature (non-conducting isothermal enclosure), the quality and the quantity of radiation inside such an enclosure depend only on its temperature and is entirely independent of both the nature of the walls and the nature of the substances which may be inside it.* This can be proved as follows :

Let us suppose two enclosures A and *B* having walls of different materials; but maintained at the same constant temperature *T*. Let each one be filled with radiant energy. Suppose that the density of radiant energy (radiant energy per unit volume) depends on the temperature as well as on the nature of the walls. Let us also suppose that the density of radiant energy is greater in A than that in *B*.

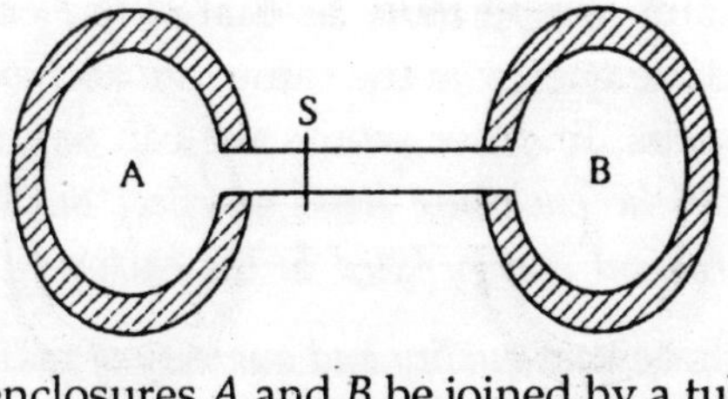

Let the enclosures *A* and *B* be joined by a tube fitted with a screen S transparent to all radiations. Evidently, more radiation will fall on the screen from *A* than from *B* and hence on the whole, there will be a flow of radiations from *A* to *B*.

After sometime, let the enclosures be separated and allowed to attain equilibrium once again with their new energy densities. The energy density in *A* is less than before and hence its walls will loose heat into the enclosure and cool down till equilibrium is again attained. On the other hand, the energy density in *B* is greater than before, therefore, its walls will absorb energy from within and get heated. This means that a difference of temperature will be produced between A and *B*, although they were originally at one temperature. Alternatively, we have created a difference of temperature between the two enclosures *A* and *B* without any expenditure of energy. Using *B* as hot body (source) of a heat engine and *A* as a cold body (sink), we can obtain work without expenditure of any energy which is contrary to the second law of thermodynamics. Therefore, our original assumption that the energy density of radiation in *A* is greater than in *B* is incorrect and hence the energy densities in *A* and *B* must be identical. Thus, we can say that *the energy density* of *the radiation in a uniform temperature enclosure depends on its temperature and is entirely independent of the nature of the walls.*

Now, let us suppose that the bodies *A*, *B* and *C* are placed inside an isothermal enclosure. Experience shows that each body will attain the temperature of the enclosure irrespective of its nature or original temperature. As the bodies *A*, *B* and *C* are at the same temperature as that of the enclosure, each body must radiate energy at the same rate and exactly of same kind as it absorbs. In other words we can say that *the bodies contained inside the enclosure have no effect on the quantity or quality of the radiant energy filled in the enclosure.*

We conclude that *quality and quantity of radiation inside an isothermal enclosure depend only on its temperature and not on the nature of the walls or on the bodies present inside it.*

Black–Body and Black–Body Radiation : A perfectly black body is one which absorbs all the heat radiations of whatever wavelength, incident on it. It neither reflects nor transmits any of the incident radiations and, therefore, appears black whatever be the colour of incident radiation.

Let a black-body be placed in an isothermal enclosure. The body will emit the full radiation of the enclosure after it is in thermal equilibrium with the enclosure. These radiations are independent of the nature of the substance. Clearly, the radiation from an isothermal enclosure is identical with that from a black-body at the same temperature. Therefore, the heat radiations in an isothermal enclosure are termed as black-body radiation.

In practice, no substance possesses strictly the properties of a black-body. Lamp–black and the Platinum black are the nearest approach to a black-body. However, the bodies showing close approximation to a perfectly black-body can be constructed.

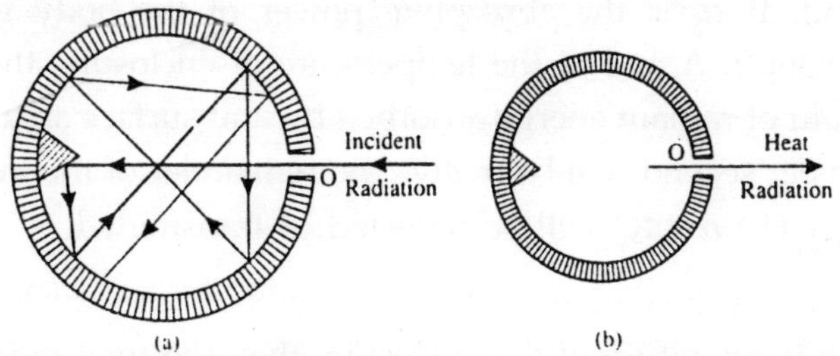

Ferry's Black-body : It consists of a hollow copper sphere blackened inside with a small fine hole *O* in the surface. When the radiations enter the hole, they suffer a number of reflections at the inner walls of the sphere until it is completely absorbed. To avoid direct reflection of the radiation from the inner surface, a pointed projection is made in front of the hole as in fig. (a). Thus the small hole acts as a black-body absorber. When this sphere is placed in a bath at a fixed temperature, the heat radiations come out of the hole as shown in fig. (b). It is to be noted that only the hole and not the walls of the sphere, acts as a black-body radiator.

Wien's Black-body : It consists of a long hollow metallic cylinder C blackened inside with heating coils around it. The cylinder is surrounded by concentric porcelain tubes. Heat radiations emerge out of the hole. The radiations from the inner surface pass through a number of limiting diaphragms. The temperature of central part of the cylinder is measured by a thermocouple. The hole acts as a black-body radiator.

Kirchhoff's Law

Statement: It states that *the ratio of the emissive power (e_λ) to the absorptive power (α_λ) for a given wavelength at a given temperature is the same for all bodies and equal to the emissive power of a perfectly black-body (E_λ) at that temperature.*

Proof : Let us consider a body placed in an isothermal enclosure. Let dQ be the amount of radiant energy of wavelength lying between λ, and $\lambda + d\lambda$, incident on unit surface area per second. If α_λ is the absorptive power of the body for the wavelength A and at the temperature of enclosure, then the amount of radiant energy absorbed by unit surface area of the body per second will be $\alpha_\lambda\, dQ$. The remainder of the incident energy $(1 - \alpha_\lambda)\, dQ$ will be reflected or transmitted.

If e_λ is the emissive power of the body for wavelength λ at the temperature of the enclosure, then the total radiations lying between wavelength λ and $\lambda + d\lambda$ emitted by unit surface area of the body per second is $e_\lambda\, d\lambda$. *As* the body is in temperature equilibrium with the enclosure, energy radiated by the body must be equal to that received by it, i.e.,

$$(1 - \alpha_\lambda)\, dQ + e_\lambda\, d\lambda = dQ$$

or $$e_\lambda\, d\lambda = a_\lambda\, dQ$$

or $$\frac{e_\lambda}{a_\lambda} = \frac{dQ}{d\lambda}. \qquad \ldots(1)$$

But $\frac{dQ}{d\lambda}$ depends only on temperature, therefore $\frac{e_\lambda}{a_\lambda}$ is the same for all substances for a given temperature, *i.e. the ratio of emissive power to the absorptive power for radiation of given wavelength is same for all bodies at a given temperature.*

If the body under consideration is perfectly black, the absorptive power $a_\lambda = 1$ for all wavelengths and e_λ has maximum value which we denote by E_λ. Therefore for such a body, we have

$$E_\lambda \, d\lambda = dQ$$

or
$$E_\lambda = \frac{dQ}{d\lambda} \quad \text{... (2)}$$

Comparing (1) and (2), we have for any body

$$\frac{e_\lambda}{a_\lambda} = E_\lambda \quad \text{... (3)}$$

i.e., the ratio of the emission power to absorptive power is equal to emissive power of a perfectly black-body at a given temperature which is Kirchhoff's law.

Importance

(i) If emissive power e_l of a body is greater, Kirchhoff's law tells that corresponding absorptive power a_λ must be greater, i.e. from Kirchhoff's law it follows that *good absorbers of radiation are also good emitters of radiation.*

(ii) Kirchhoff's law established that the atoms of each element give a spectrum which is characteristic of that element. Thus, the presence of characteristic spectral lines in the spectrum of an element indicates the presence of the element in the substance of which spectrum is taken.

Important Applications

(i) If a polished metal ball having a black spot on its surface is heated to a high temperature, the *black spot shines more brightly* than the polished surface. Its reason is that the black spot is a better absorber of radiations, and hence according to Kirchhoff's law, it will be a better emitter.

(ii) A *red piece of glass glows with a green light when heated.* Actually red glass appears red because it reflects red light and absorbs rest from the white light. The effect of visible spectrum minus red is green. Hence, red glass emits green on heating which it has absorbed when it was cold.

Similarly, green glass appears red on heating strongly.

(iii) If sodium is heated, its emission spectrum consists of two yellow lines (D_1 and D_2). On the other hand, when white light is passed through a cooler sodium vapour, the continuous spectrum of white light slows two dark lines in exactly the positions of D_1 and D_2 lines of emission spectrum. It shows that *sodium vapour which emits two yellow lines strongly, is also a good absorber of light of these two wavelengths.*

(iv) The best example of Kirchhoff's law is the *Fraunhoffer dark lines* found in the continuous spectrum of the sun. Actually, sun consists of a central glowing mass which emits the continuous spectrum. When radiations from the central mass pass through the cooler surrounding atmosphere, containing various elements like hydrogen, nitrogen, sodium, copper etc. in the gaseous state, these elements absorb these wavelengths which they emit at a higher temperature. Hence, these wavelengths are absent in the solar spectrum and we see dark lines in their places, which are known as Fraunhoffer lines.

Pressure of Radiation

As we know that the radiation possesses the properties of light, like light, it exerts a small but definite pressure on the surface on which it is incident. Maxwell proved on the basis of his electromagnetic theory that the pressure is equal to the energy density (i.e. the amount of radiation per unit volume) for normal incidence on a surface. Here, we shall deduce the expression for the pressure of radiation on the basis of quantum theory.

Let a photon of energy $h\nu$ move with velocity of light c. According to the theory of relativity, this energy of the photon corresponds to a mass m given by

$$mc^2 = h\nu$$

$$m = \frac{h\nu}{c^2}.$$

Then the momentum of photon = mass × velocity

$$= \frac{h\nu}{c^2} c = \frac{h\nu}{c}$$

$$= \frac{e}{c}, e \text{ being energy of photon.}$$

Now the momentum incident on the surface PQ per unit area per second is given by

$$P = \Sigma \frac{e}{c} = \frac{\Sigma e}{c} \quad \text{... (1)}$$

but Σe = total energy incident on the surface per unit area per second = E (say).

Substituting this in (1), we get

$$P = \frac{E}{c}.$$

If u is the energy density, i.e. energy per unit volume, then the total energy passing through any area s of the surface normal to the radiation per second = $u \times$ sc.

Therefore the energy flux, i.e., the energy radiations per unit area per second,

$$E = \frac{ucs}{s} = uc. \qquad \text{... (2)}$$

Substituting this in (1), we get

$$p = \frac{uc}{c} = u$$

If all of this momentum, i.e., the momentum per unit area per second, is absorbed, it will give the rate of change of momentum per unit area, i.e. the pressure of radiation.

$$p = u. \qquad \text{... (3)}$$

Thus, for normal incidence on a surface, *the pressure of radiation is equal to the energy density.*

Pressure of Diffuse Radiation

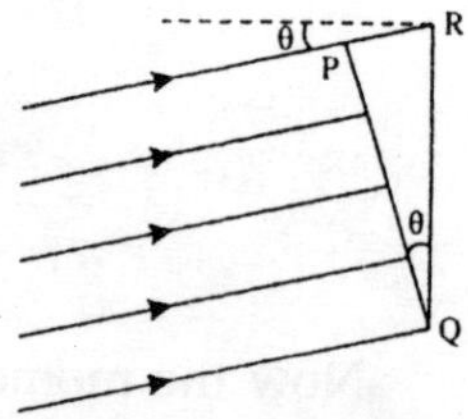

If the radiation is diffuse, i.e., if the photons move in all possible directions, the pressure of radiations will be one third of the above which may be found as follows :

Let the beam of radiation make an angle θ with the normal to the surface *QR*. The energy incident on the surface *PQ* per unit area per second = *uc*.

If S is the surface area of the plane *PQ*, the energy incident on the surface *PQ* per second = *uc* . S.

But surface area of the plane *PQ*

= (surface area of plane *QR*) × cos θ

= S′ cosθ

where S′ = surface area of plane *QR*.

∴ The energy incident on the surface *PQ* per second

= *uc*. *S*′ cosθ.

From fig., it is obvious that the total energy crossing the plane *PQ* per second is equal to the energy incident on the plane *QR* per second.

Therefore, the energy incident on the plane QR per second

$$= uc\ S'\cos\theta.$$

So that the energy incident on the plane QR per unit area per second

$$= \frac{ucS'\cos\theta}{\text{Surface area of plane } QR}$$

$$= \frac{ucS'\cos\theta}{S'} = uc\cos\theta$$

Therefore, the momentum incident on the surface QR per unit area per second $= \dfrac{uc'S\cos\theta}{c} = u\cos\theta.$

The component of this momentum in a direction normal to the surface

$$QR = (u\cos\theta)\cos\theta$$

$$= u\cos^2\theta. \qquad \text{... (4)}$$

If all this momentum is absorbed, then eqn. (4) gives the total rate of change of momentum per second per unit area normal to the surface QR, *i.e.*, the pressure of radiation on the surface QR is given by

$$p = u\cos^2\theta. \qquad \text{... (5)}$$

In the case of diffuse radiations, the radiation is incident from all possible directions with equal probability. To find the pressure of diffuse radiation, we must take average value of $\cos^2\theta$ over all directions. Hence pressure of diffuse radiation is given by

$$p = u\,\overline{\cos^2\theta} \qquad \text{... (6)}$$

where $\overline{\cos^2\theta}$ represents averaged $\cos^2\theta$ over all directions.

The average value of $\cos^2\theta$ over all direction is 1/3. Which may be seen as follows :

Consider the radiation to be equivalent to a large number of beams of equal intensity, say N, distributed uniformly over all directions.

In order to calculate $u \cos^2 \theta$ imagine a hemisphere of radius r round the centre O of the element RQ. Now from the surface of the hemisphere cut out a ring shaped element ($ABCD$) by means of two cones of semi angles θ and $\theta + d\theta$ drawn from O as apex and OM (which is perpendicular to RQ) as axis. The area of this ring $= 2\pi r \sin \theta \, . \, r \, d\theta = 2\pi r^2 \sin \theta \, d\theta$.

As N beams are considered to be uniformly distributed over the surface of the hemisphere which is equal to $2\pi r^2$.

The number of beams passing through the area of the ring.

$$dN = \frac{2\pi \, r^2 \sin \, \theta \, d \, \theta}{2\pi r^2} N = N \sin \theta \, d \, \theta. \quad \text{... (7)}$$

Fraction of beams incident between angle θ and $\theta + d\theta$

$$= \frac{dN}{N} = \sin \theta \, d\theta$$

The pressure p can, therefore, be written as

$$P = \frac{dN}{N} \times u \cos^2 \theta = \sin \, \theta \, d\theta \, . \, u \cos^2 \theta$$

$$= u \cos^2 \theta \sin \, \theta \, d\theta. \quad \text{... (8)}$$

For all possible rings within the hemisphere, θ may vary from 0 to $\frac{\pi}{2}$, therefore eqn. (8) in the integration form may be written as

$$p = \int_0^{\pi/2} u \, \cos^2 \theta \sin \theta \, d\theta$$

$$= u \int_0^{\pi/2} \cos^2 \theta \sin \theta \, d\theta \quad \text{... (9)}$$

Now put $\cos 0 = x$

$\therefore \quad - \sin \, \theta \, d\theta = dx$

$$\therefore \quad p = u\int_1^0 x^2(-dx) = u\int_0^1 x^2 dx$$

$$= u\left[\frac{x^2}{3}\right]_0^1 = \frac{u}{3}.$$

$\therefore$ Pressure of diffuse radiation $= \frac{1}{3} \times$ energy density.

PROBLEMS

1. *The order to magnitude of the energy received from sun at earth's surface is 10^{-1} joule cm^{-2} sec^{-1}. Calculate the order of magnitude of the total force due to solar radiation on the earth (assumed perfectly absorbing). Given earth's diameter $\approx$ 10^{-7} meter.*

Solution : The energy received on earth's surface per unit area per sec.

$$E = 10^{-1} \text{ joule/cm}^2 \text{ sec.} = 10^3 \text{ joule m}^{-2} \text{ sec.}^{-1}$$

$\therefore$ The radiation pressure,

$$p = \frac{E}{c} = \frac{10^3}{3\times10^8} N/m^2 = 3.33\times10^{-6} N/m^2$$

$\therefore$ Total force due to solar radiation on the earth

$$= p \times \text{surface area}$$

$$= 3.33 \times 10^{-6} \times 4\pi r^2$$

$$= 3.33\times10^{-6}\times4\times3.14\times\left(\frac{10^7}{2}\right)^2$$

$$= 1.05 \times 10^9 \text{ N}.$$

Stefan-Boltzmann Law

Stefan's law states that *the rate of emission of radiant energy by unit area of perfectly black-body is directly proportional to the fourth power of its absolute temperature.* In symbols

$$E \propto T^4$$

or
$$E = \sigma T^4$$

where σ *is* a constant and is called Stefan's constant.

The law in the above form refers to the emission only and not to the net loss of heat by body after exchange with the surroundings. It simply deals with the amount of heat emitted by the body by virtue of its temperature irrespective of what it receives from the surroundings. The law can be extended to represent the net loss of heat and may be enunciated as follows:

A black-body at absolute temperature T surrounded by another black-body at absolute temperature T_0 not only loses an amount of energy σT^4 but also gains σT_0^4, thus the amount of heat lost by the former per unit time is given by

$$E = \sigma (T^4 - T_0^4)$$

The law is known as Stefan Boltzmann's law as Boltzmann had deduced it thermodynamically.

Thermodynamical Proof : Let us consider a cylindrical enclosure $ABCD$ of uniform cross-section with perfectly reflecting walls and provided with e perfectly reflecting piston P. Let it be filled with diffuse radiations of density u at uniform temperature T.

If V be the volume of the enclosure; then the total internal energy of radiations inside the enclosure is given by

$$U = uV. \qquad \text{... (1)}$$

Let a small amount of heat dQ flow in the enclosure from outside and at the same time the piston moves out so that the volume changes by a small amount dV. In this process the temperature and as a result the energy density u changes by an infinitesimal amount. If dU is the change in internal energy of radiation and dW is the external work done by radiation in the expansion of volume by an amount dV, then by the first law of thermodynamics

$$dQ = dU + dW.$$

But $U = uV$ and $dW = pdV$, p being the pressure of radiation

$$dQ = d\ (dV) + p\ dV.$$

But the pressure of diffuse radiation $p = \frac{u}{3}$, therefore,

$$dQ = d(uV) + \frac{1}{3} u\ dV$$

$$= u\ dV + Vdu + \frac{1}{3} u\ dV$$

$$= V\ du + \frac{4}{3} + u\ dV. \qquad \text{... (2)}$$

If dS is the change in entropy of the radiation, then from second law of thermodynamics, we have

$$dS = \frac{dQ}{T}. \qquad \text{... (3)}$$

Substituting value of dQ from (2) in (3), we have

$$dS = \frac{V}{T} du + \frac{4u}{3T} dV. \qquad \text{... (4)}$$

From this expression it is clear that S is function of two independent variables u and V, i.e.,

$$S = f\ (u, V)$$

$$dS = \frac{\partial S}{\partial u} du + \frac{\partial S}{\partial V} dV \qquad \text{... (5)}$$

Comparing (4) and (5), we get

$$\frac{\partial S}{\partial u} = \frac{V}{T} \text{ and } \frac{\partial S}{\partial u} = \frac{4u}{3T}. \qquad \text{... (6)}$$

As dS *is* a perfect differential, we have

$$\frac{\partial^2 S}{\partial u\, \partial V} = \frac{\partial^2 S}{\partial V\, \partial u}$$

$$\frac{\partial}{\partial u}\left(\frac{\partial S}{\partial V}\right) = \frac{\partial}{\partial V}\left(\frac{\partial S}{\partial u}\right). \qquad \text{... (7)}$$

Substituting the values of $\frac{\partial S}{\partial V}$ and $\frac{\partial S}{\partial u}$ as from (6) in (7), we get

$$\frac{\partial}{\partial u}\left(\frac{4u}{3T}\right) = \frac{\partial}{\partial V}\left(\frac{V}{T}\right). \quad \text{... (8)}$$

As T is independent of V and is a function of u only, on differentiating (8), we get

$$\frac{4}{3}\cdot\frac{1}{T} - \frac{4}{3}\cdot\frac{u}{T^2}\cdot\frac{\partial T}{\partial u} = \frac{1}{T}$$

or $$\frac{1}{3}\cdot\frac{1}{T} = \frac{4}{3}\cdot\frac{u}{T^2}\cdot\frac{\partial T}{\partial u}$$

or $$\frac{\partial u}{u} = 4\,\frac{\partial T}{T}.$$

Integrating, we have

$$\log u = 4 \log T + \log A,$$

log A being constant of integration

which gives $$u = AT^4. \quad \text{... (9)}$$

Now it can be shown that the total rate of emission of radiant energy per unit area is related to the energy density by the relation

$$E = \frac{1}{4}uc$$

where c is the velocity of light.

$$\therefore \quad E = \frac{1}{4}AcT^4 \text{ or } E = \sigma T^4. \quad \text{... (10)}$$

where σ is a constant called Stefan's constant. The value of Stefan's constant in C.G.S. system is 5.672×10^{-5} erg. cm^{-2} sec^{-1} $(K)^{-4}$ and in M.K.S. system it is 5.672×10^{-8} joule m^{-2} sec^{-1} $(K)^{-4}$.

[*Note :* For low temperature difference, Newton's law of cooling is just an approximation of Stefan Boltzmann law.

If $T - T_0 = \Delta T$

or $T = T_0 + \Delta T$

Then from Stefan Boltzmann law

$$E = \sigma\left[(T_0 + \Delta T)^4 - T_0^4\right]$$

$$= \sigma T_0^4\left[\left(1 + \frac{\Delta T}{T}\right)^4 - 1\right]$$

$$= \sigma T_0^4\left[1 + 4\frac{\Delta T}{T_0} - I\right] = 4\sigma T_0^3 \Delta T$$

$\therefore$ $E \propto \Delta T$

which is Newton's law of cooling.]

Determination of Stefan's Constants : The apparatus used for determination of Stefan's constant consists of a hollow metallic hemisphere H blackened inside and placed in a wooden box W, lined with tin which serves as a steam chamber. Steam can be passed in the chamber when desired and then in steady state the temperature of hemisphere remains constant, equal to that of steam and may be measured by thermometers T_1 and T_2. The hemisphere H rests symmetrically on a platform AB which has a small hole at its centre. A small silver disc D blackened at its top surface can be fitted or taken out from the hole. One junction of the silver–constantan couple is soldered to the lower surface of D; while the other is placed in the water bath or sand container C. A sensitive galvanometer (G) is introduced in thermocouple circuit with resistance R in series.

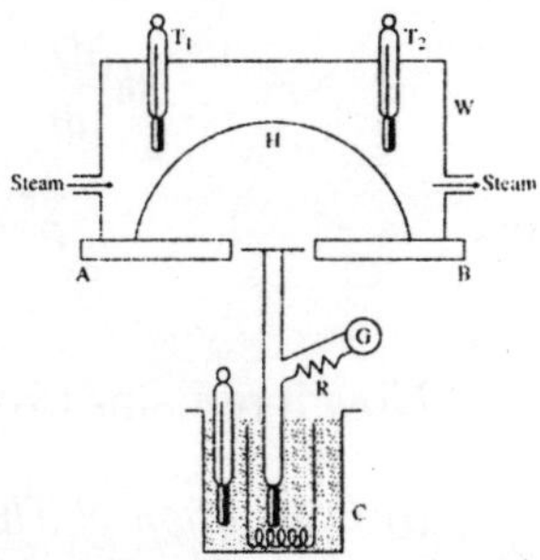

Theory: When the inner surface of H is heated by passing steam, it acts as a black body radiator. The disc D absorbs the radiation emitted by H and its temperature rises continuously; thus causing a difference of temperature in the two junctions of the thermocouple.

If T_1 is the steady state temperature of H and T_0 is that of disc D, when it is just exposed to radiation from H.

Then by Stefan–Boltzmann law of the net energy gained by the disc per second = EA

$$= \sigma\,(T_1^4 - T_0^4)\,A \text{ Joule, A belong area of disc}$$

$$= \frac{\sigma\left(T_1^4 - T_0^4\right)A}{J} \text{kilocal.}$$

If m is the mass of disc, c its specific heat in kilo cal. (°C)$^{-1}$ (kg)$^{-1}$ and $\frac{dT}{dt}$ its rate of rise of temperature at T_0, then the energy gained by the disc per sec. would be $mc\,\frac{dT}{dt}$; hence we must have

$$mc\frac{dT}{dt} = \frac{\sigma\left(T_1^4 - T_0^4\right)A}{J}$$

or
$$\sigma = \frac{Jmc}{A\left(T_1^4 - T_0^4\right)} \cdot \frac{dT}{dt} \text{ watt m}^{-2}K^{-4}. \qquad \text{... (1)}$$

Now it remains to evaluate $\frac{dT}{dt}$; which is done as follows :

(i) *Calibration of Thermo-couple* : Before passing steam in the chamber, the silver disc is kept in position in the hole. The water or sand bath (C) is heated to a suitable temperature. Due to temperature difference of junctions of thermocouple the galvanometer gives deflection. The resistance R is so seen that the deflection does not go beyond the scale of galvanometer. Now the deflection x of the galvanometer is noted for different temperatures T of the bath, when falling. The graph is plotted between temperature T on Y axis and deflection x on X-axis. The graph is in the form of a straight line whose slope (tan θ_1) gives $\frac{dT}{dt}$. During this part the disc acts as a cold junction of thermocouple.

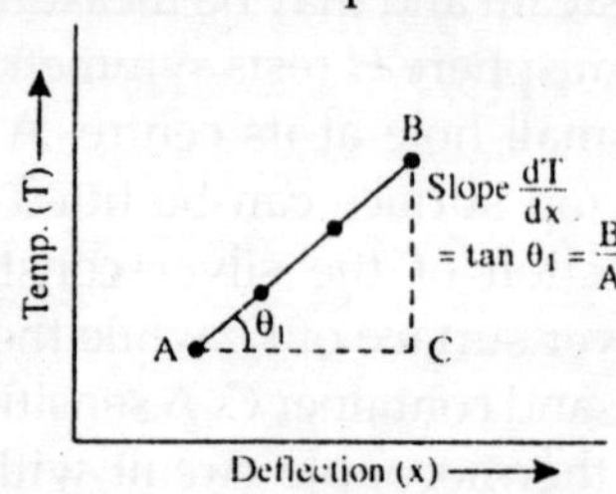

(ii) *Determination of* $\frac{dT}{dt}$: The disc is removed from the hole and steam is passed through chamber *W*. When temperature of chamber becomes steady, the disc is again positioned in the hole. It now absorbs radiation from chamber and so its temperature rises.

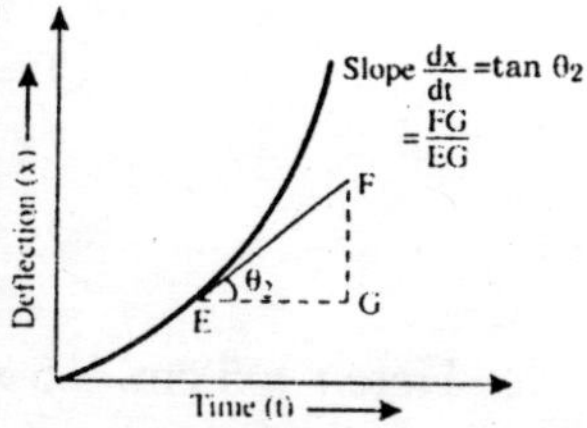

Therefore, deflection in galvanometer increases. A graph is plotted between deflection *x* and time *t*. A point *E* near the origin is chosen and a tangent is drawn from *E* on the graph.

The slope of tangent (tan θ_2) gives $\frac{dx}{dt}$.

$$\text{Thus } \frac{dT}{dt} = \frac{dT}{dx} \times \frac{dx}{dt} = \tan\theta_1 \times \tan\theta_2 = \frac{BC}{AC} \times \frac{FG}{EG}.$$

Thus the value of $\frac{dT}{dt}$ at temperature corresponding to point E is obtained knowing constants *M*, *c* and *A* of disc, temperature *T*, of steam chamber, T_0 the room temperature; σ may be evaluated from (1).

PROBLEMS

1. ***Calculate the maximum amount of heat which may be lost per second by radiation from a sphere of 10 cm in diameter at a temperature of 227°C when placed in an enclosure at a temperature of 27°C.***

$$[\sigma = 5.7 \times 10^{-12} \textit{ watts} \times cm^{-2} (°C)^{-4}].$$

Solution: The rate of loss of energy will be maximum if the sphere radiates as black body. By Stefan Boltzmann's law, the net energy lost per unit area of the sphere per second is given by

$$E = \sigma\,(T^4 - T_0^4).$$

where *T* and T_0 are the absolute temperature of sphere and the surroundings respectively and σ is Stefan's constant.

Total surface area of sphere = $4\pi\, r^2$.

$\therefore$ The net rate of loss of heat from the total surface of the sphere

$$Q = E \times 4\pi r^2$$

$$Q = \sigma\,(T^4 - T_0^4) \times 4\pi r^2$$

$$= \frac{\sigma\left(T^4 - T_0^4\right) \times 4\pi\, r^2}{J} \text{ calories/second}$$

Here $r = 5$ cm. $= 5 \times 10^{-2}$ m, $T = 227^\circ\text{C} = (227+273)$ K$=500$ K

$T_0 = 27^\circ\text{C} = (27 + 273)$ K $= 300$ K

$\sigma = 5.7 \times 10^{-12}$ watt cm.^{2} $(\text{K})^{-4} = 5.7 \times 10^{-8}$ watt m^{-2} $(\text{K})^{-4}$

and $J = 4.2$ Joule/cal.

$$\therefore \quad Q = \frac{5.7 \times 10^{-8}\left[(500)^4 - (300)^4\right] \times 4 \times 3.14 \times \left(5 \times 10^{-2}\right)^2}{4.2}$$

$= 23.3$ cal/s.

2. *Luminosity of Rigel Star in Orion constellation is 17000 times that of our sun. If the surface temperature of the sun is 6000 K, calculate the temperature of the star.*

Solution : From Stefan Boltzmann's law the energy radiated per unit surface of the body is given by

$$E = \sigma\,(T^4 - T_0^4)$$

If the temperature of the surroundings T_0 is very-very less than that of body (as in this case), we have

$$E = \sigma\, T^4.$$

Therefore, if E_1 and E_2 represent the luminosities and T_1 and T_2 the absolute temperatures of the star and the sun respectively, then we have

$$E_1 = \sigma\, T_1^4$$

$$E_2 = \sigma\, T_2^4$$

$$\frac{E_1}{E_2} = \frac{T_1^4}{T_2^4}$$

given $$\frac{E_1}{E_2} = 17000, T_2 = 6000K$$

$$17000 = \frac{T_1^4}{(6000)^4}$$

or $$T_1^4 = (6000)^4 \times 17000$$

or $$T_1 = 6000 \times (17000)^{1/4}$$
$$= 68520 \text{ K}.$$

3. ***Two ideal black bodies A and B at temperatures 227°C and 327°C respectively are placed in an evacuated enclosure whose walls are blackened and kept at 27°C. Compare their rates of loss of heat.***

Solution: From Stefan Boltzmann's law the energy lost by any body per second per m² is

$$E = \sigma (T^4 - T_0^4)$$

For black body A, $T = 227°\text{C} = 500 \text{ K}$

and $$T_0 = 27°\text{C} = 300 \text{ K}$$

$$\therefore \quad E_1 = \sigma [(500)^4 - (300)^4]$$
$$= 544 \times 10^8 \sigma.$$

For black body B, $T = 327°\text{C} = 600 \text{ K}$

and $$T_0 = 27°\text{C} = 300 \text{ K}$$

$$\therefore \quad E_2 = \sigma [(600)^4 - (300)^4]$$
$$= 1215 \times 10^8 \sigma$$

$$\frac{E_1}{E_2} = \frac{544 \times 10^8 \sigma}{1215 \times 10^8 \sigma} = \frac{544}{1215}$$

∴ Ratio of rates of *loss of A : B*

= 544: 1215.

4. ***If each square cm. of sun's surface radiates energy at the rate of 1.5 × 10³ cal. sec.⁻¹ cm⁻² and Stefan's constant is 5.7 ×***

10^{-8} joule sec.$^{-1}$ m^{-2} (K)$^{-4}$, calculate the temperature of the sun's surface in degree Celsius.

Solution: From Stefan–Boltzmann's law

$$E = \sigma\,(T^4 - T_0^4)$$

If $T >> T_0$ (as in this case) then we have

$$E = \sigma\, T^4 \qquad \text{... (i)}$$

The energy radiated by sun

$$E = 1.5 \times 10^3 \text{ cal. cm.}^{-2}\text{ s}^{-1}$$

$$= 1.5 \times 10^3 \times 4.2 \text{ joule } (10^{-2}\text{ m.})^{-2}\text{ s}^{-1}$$

$$= 1.5 \times 4.2 \times 10^7 \text{ joule m}^{-2}\text{ s}^{-1}$$

and $$\sigma = 5.7 \times 10^{-8} \text{ joule s}^{-1}\text{ m.}^{-2}\text{ (K)}^{-4}$$

$\therefore$ Substituting these values in eqn. (i), we get

$$1.5 \times 4.2 \times 10^7 = 5.7 \times 10^{-8}\, T^4$$

or $$T^4 = \frac{1.5 \times 4.2 \times 10^{-7}}{5.7 \times 10^{-8}}$$

$\therefore$ $$T = \left[\frac{1.5 \times 4.2}{5.7} \times 10^{15}\right]^{1/4} = 5764 \text{ K}$$

$$= 5492°\text{C}.$$

5. ***A body of mass 10 gram is kept in an enclosure of temperature 27°C. If the temperature of the body is 127°C, its specific heat is 0.1 kilo calorie per kg per degree C and area of emitting surface of the body is 10^{-3} m^2, find out the rate of cooling of the body.***

[σ = 5.72 × 10^{-8} *joule m.*$^{-2}$ s^{-1} (°C)$^{-4}$].

Solution : The energy emitted by a body of surface area A at temperature *TK* in time *t* is

$$Q = A\sigma\,(T^4 - T_0^4)t$$

$\therefore$ $$\frac{dQ}{dt} = A\sigma\left(T^4 - T_0^4\right)$$

But $Q = msT$ for body mass m and specific heat *s*.

$$\therefore \quad \frac{d(msT)}{dt} = A\sigma\left(T^4 - T_0^4\right)$$

$\therefore$ Rate of cooling of the body,

$$\frac{dT}{dt} = \frac{A\sigma}{ms}.\left(T^4 - T_0^4\right)$$

$$= \frac{\left(10^{-2}m^2\right)\left(\frac{5.72\times10^{-8}}{4.2\times10^3}\text{ Kilocal.m}^{-2}s^{-1}\left({}^\circ\text{C}\right)^{-4}\right)\left(400^4 - 300^4\right)\left({}^\circ\text{C}\right)^4}{\left(10\times10^{-3}\text{kg.}\right)\left\{\left(0.1\text{ Kilocal.kg.}^{-1}\left({}^\circ\text{C}\right)^{-1}\right\}\right.}$$

$$= \frac{10^{-3}\times5.72\times10^{-8}\times\left(400^4 - 300^4\right)}{10\times10^3\times0.1\times4.2\times10^3}\ {}^\circ\text{C/s.}$$

$= 0.23°\text{C s}^{-1} = 0.23\text{ K s}^{-1}$.

6. *A black body with an initial temperature of 300°C is allowed to cool inside an evacuated enclosure surrounded by melting ice at the rate of 0.35°C per second. If the mass, specific heat and surface area of the body are 32 g 0.10 cal hg / °C and 8 cm [2] respectively, calculate the Stefan's constant.*

Solution: According to Stefan Boltzmann's law, the energy radiated per second per unit area from a body at absolute temperature *T*, surrounded by an enclosure of absolute temperature T_0, is given by

$$E = \sigma\ (T^4 - T_0^4)$$

If *A* is the surface area of the body, the loss of energy radiated by the whole body per second

$$Q = \sigma\ (T^4 - T_0^4)\ A\ \text{joule/sec.}$$

$$= \frac{\sigma\left(T^4 - T_0^2\right)A}{J}\ \text{Kilo cal/sec.} \qquad \text{... (i)}$$

where $J = 4.2 \times 10^{-3}$ joule/kilo cal.

The energy radiated by the body per sec. is also given by

$$Q = \text{mass} \times \text{specific heat} \times \text{fall in temp. per sec.}$$

$$= 32 \times 10^{-3} \text{ kg.} \times 0.10 \text{ kilo cal./kg °C} \times (0.35\text{°C})$$

$$= 1.12 \times 10^{-3} \text{ kilo cal. /sec.} \qquad \text{... (ii)}$$

From (i) and (ii), we have

$$\frac{\sigma\left(T^4 - T_0^4\right)A}{J} = 1.12 \times 10^{-3}$$

or

$$\sigma = \frac{1.12 \times 10^{-3} \times J}{\left(T^4 - T_0^4\right)A}$$

$$= \frac{1.12 \times 10^{-3} \times 4.2 \times 10^3}{\left[(573)^4 - (273)^4\right] \times 8}$$

$$= 5.7 \times 10^{-8} \text{ joule m}^{-2} \text{ sec.}^{-1} \text{ (K)}^{-4}.$$

7. Calculate the energy radiated per minute from the filament of an incandescent lamp at 2000 K if the surface area is 5 × 10^{-5} meter^2 and its relative emittance is 0.85.

[Given σ = 5.7 × 10^{-8} M.K.S. units].

Solution: According to Stefan-Boltzmann's law, the energy radiated by a body per unit area per second placed in an enclosure, is given by

$$E = \sigma\,(T^4 - T_0^4) = \sigma\,T^4 \text{ if } T >> T_0$$

where T is the temperature of the body and T_0 is the temperature of the enclosure.

If the body has an emissivity, (i.e. relative emittance) e, then

$$E = \sigma\, e\, T^4.$$

The total energy radiated by the body per second

$$= \sigma e\,(T^4 - T_0^4) \times \text{surface area}$$

$$= 5.7 \times 10^{-8} \times 0.85 \times (2000)^4 \times (5 \times 10^{-5}).$$

∴ The energy radiated by the body per minute

$$= 5.7 \times 10^{-8} \times 0.85 \times (2000)^4 \times (5 \times 10^{-5}) \times 60$$

$$= 2315 \text{ joule.}$$

Distribution of Energy in The Spectrum of a Blackbody Radiation : Lummer and Pringsheim investigated the distribution of energy among the radiations emitted by a black-body at different temperatures.

The experimental arrangement of Lummer and Pringsheim is shown in fig. Their black-body was an electrically heated chamber with a small aperture whose temperature was measured by a thermocouple.

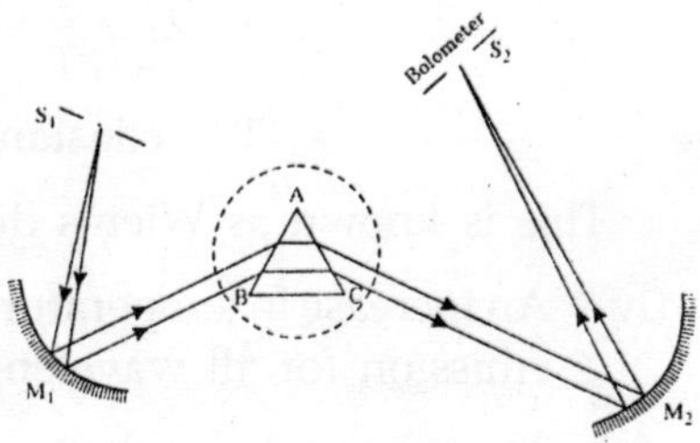

The radiations from the black body pass through the slit S_1 and fall on the reflector M_1. After being reflected, the parallel beams of radiation fall on a rock-salt or fluorspar prism ABC placed on the turn table of the spectrometer. The emergent light is focussed by the reflector (concave mirror) M_2 on a line bolometer placed behind the slit S_2. The bolometer is connected to a sensitive galvanometer. The turn table is rotated slowly so that different parts of the radiation spectrum successively fall on the bolometer and the corresponding deflections in galvanometer connected in the bolometer circuit are read. The intensity of each line is proportional to the deflection in the galvanometer. Then a curve (fig.) is drawn between intensity and the wavelength.

Then the body is heated to different temperatures and the curves are drawn for various temperatures.

Result : The curves show :

(i) The energy is not uniformly distributed in the radiation spectrum of a black-body.

(ii) At a given temperature, the intensity of radiations

increases with increase of wavelength and becomes maximum at a particular wavelength. By further increasing the wavelength, intensity of heat radiations decreases.

(iii) An increase in temperature causes a decrease in λ_m where λ_m is the wavelength for which the energy emitted is maximum. Thus λ_m is inversely proportional to absolute temperature i.e.,

$$\lambda_m \propto \frac{1}{T}$$

or $$\lambda_m T = \text{constant.}$$

This is known as Wien's displacement law.

(iv) An increase in temperature causes an increase in energy emission for all wavelengths.

(v) The area under each curve represents the total energy emitted by the body at a particular temperature for the range of the wavelengths considered. This area increases with increase in temperature. It is found that area is directly proportional to the fourth power of absolute temperature i.e. $E \propto T^4$ which is Stefan's law.

Wien's Displacement Law

It states that *the product of the wavelength corresponding to maximum energy* λ_m *and the absolute temperature T is constant i.e.*

$$\bar{\lambda}_m T = \text{constant.}$$

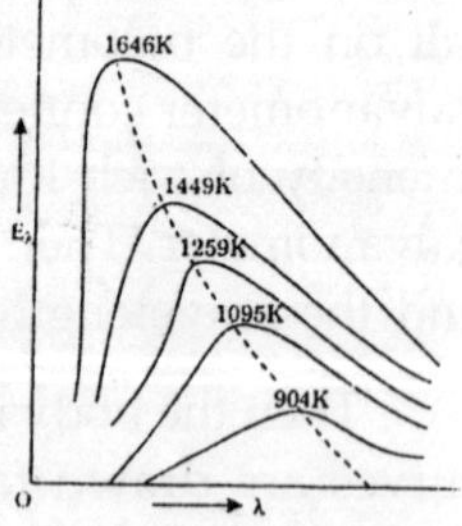

This constant is called Wien's displacement constant and has a value 0.2896 cm K = 0.2896 × 10^{-2} mK

According to this law λ_m decreases with increase in temperature as we have already shown in previous fig.

Deduction : Let us imagine a spherical enclosure of perfectly reflecting walls and capable of expanding. Let it be filled with diffuse radiations of energy density u at a uniform temperature T.

If V is the volume of the enclosure, then total internal energy of radiations is given by

$$U = uV$$

Let us now suppose that the walls of the enclosure move outward slowly with uniform velocity so that the radiation inside it expands adiabatically. If dV is the change in volume of the enclosure, the work done by the pressure of radiation on the walls of the enclosure = pdV which is drawn from internal energy of radiation. If dU is the decrease in internal energy, then from first law of thermodynamics, we have

$$dU + pdV = dQ = O \qquad \text{...(1)}$$

Since $dQ = 0$ as the change is adiabatic

We know $\quad p = \frac{1}{3}u.$

Substituting values of p and U in (1), we get

$$d(uV) + \frac{1}{3}u\,dV = 0$$

or $\quad udV + Vdu + \frac{1}{3}u\,dV = 0$

or $\quad \frac{4}{3}u\,dV = -V\,du$

or $\quad \frac{4}{3} \cdot \frac{dV}{V} = -\frac{du}{u}. \qquad \text{... (2)}$

Integrating we get

$$\frac{4}{3}\log V = -\log u + \text{constant}$$

or $\quad \log V^{4/3} + \log u = \text{constant}.$

or $\quad \log V^{4/3} u = \text{constant}.$

But $u = AT^4$, where A is constant.

$\therefore$ $$V^{4/3}\, T^4 = \text{constant}.$$

or $$V^{1/3}\, T = \text{constant}. \quad \text{... (3)}$$

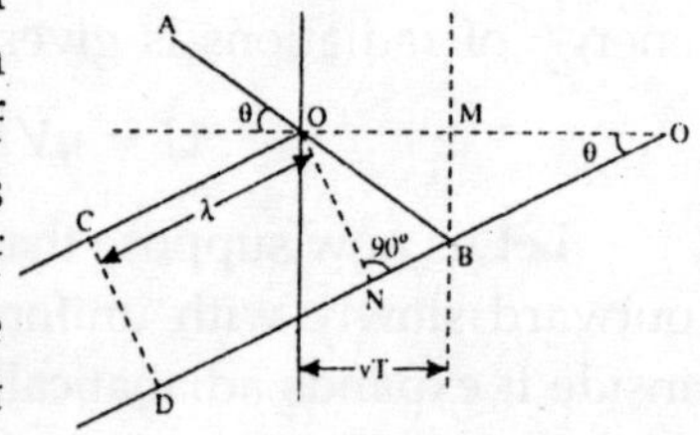

Let us now find a change in wavelength suffered by a ray on reflection at the moving walls of the enclosure due to Doppler's effect. Let AO be a ray of wavelength λ incident at an angle θ on the wall as shown in fig. Let a particular wave crest strike the wall at O. The crest will be reflected along OC. If $OC = \lambda$ then as the reflected wave crest reaches C, the next will reach at O in time T, T being time period of wave motion. During this time the wall has traversed a distance $OM = vT$. Therefore path followed by the second wave crest will be OBD. Then increase in wavelength due to one reflection is

$$= OB + BN = O'B + BN$$
$$= O'N = OO' \cos\theta$$
$$= 2OM \cos\theta.$$
$$= 2vT \cos\theta.$$
$$= \frac{2v\lambda}{c}\cos\theta \quad \text{... (4)}$$

Since $T = \dfrac{\lambda}{c}$, c being velocity of radiation.

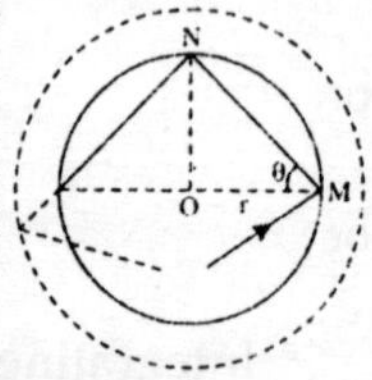

In order to calculate the number of reflections per second, let us consider a ray striking the wall of the sphere at M (Fig.) and reflected along MN making an angle θ with the normal.

This distance traversed by the beam before it suffers the next reflection $= MN = 2r\cos\theta$.

Therefore, the time between two successive reflections

$$= \frac{2r\cos\theta}{c}.$$

$\therefore$ The number of reflections per sec.

$$= \frac{c}{2r \cos \theta}.$$

If δr is the increase in radius in time δt, then the number of reflections in time δt.

$$= \frac{c}{2r \cos \theta} \delta t$$

$$= \frac{c}{2r \cos \theta} . \frac{\delta r}{v}. \qquad \text{... (5)}$$

The change in wavelength in time δt

$\delta \lambda$ = change in wavelength in one reflection × number of reflections in time δt

$$\therefore \quad \delta r = \frac{2v\lambda}{c} \cos \theta \times \frac{c}{2r \cos \theta} . \frac{\delta r}{v} \qquad \text{using (4) and (5)}$$

$$= \lambda \frac{\delta r}{r}$$

$$\therefore \quad \frac{\delta \lambda}{\lambda} = \frac{\delta r}{r}. \qquad \text{... (6)}$$

We know that volume of sphere $V = \frac{4}{3} \pi r^3$, so that the change in volume of sphere, $\delta V = \frac{4}{3} \pi . 3r^2 \delta r$

$$= \frac{4}{3} \pi r^2 . \frac{3}{2} \delta r$$

$$V = \frac{3}{r} \delta r$$

or

$$\frac{\delta r}{r} = \frac{1}{3} . \frac{\delta V}{V} \qquad \text{... (7)}$$

Substituting this in eqn. (6), we get

$$\frac{\delta \lambda}{\lambda} = \frac{1}{3} . \frac{\delta V}{V} \qquad \text{... (8)}$$

Integrating, we get $\log \lambda = \frac{1}{3} \log V + \log K$, where $\log K$ is a constant which gives

$$\lambda = KV^{1/3}$$

or $$V^{1/3} = \frac{\lambda}{K}. \quad \text{... (9)}$$

Substituting this in eqn. (3), we get

$$\frac{\lambda}{K} . T = \text{Constant}.$$

or $$\lambda T = \text{Constant}. \quad \text{... (10)}$$

Which is the usual form of Wien's displacement law. According to it if the radiation of a particular wavelength at a certain temperature is adiabatically altered to another wavelength, the temperature changes inversely.

Let us now suppose that the radiations lying between wavelength λ and λ + *d*λ are isolated in the enclosure. Let u_λ d_λ be the energy density of these radiations.

The total internal energy of these radiations $U = u_\lambda\ d_\lambda . V$. When these radiations are expanded adiabatically alone, the work done in adiabatic expansion

$$= \text{P}\ \Delta V$$

$$= \frac{1}{3} u_\lambda\ d\lambda\ \Delta V$$

∴ From 1st law of thermodynamics, we have

$$\Delta\left(u_\lambda d\lambda . V\right) + \frac{1}{3} u\lambda\ d\lambda\ \Delta V = dQ = 0.$$

(since under adiabatic conditions $dQ = 0$)

or $$\Delta u_\lambda . d\lambda . V + u_\lambda\ \Delta d\ \lambda V + u_\lambda\ d\lambda . \Delta V + \frac{1}{3} u_\lambda . d\lambda . \Delta V = 0$$

Dividing throughout by $u_\lambda . d\lambda\ V$, we get

$$\frac{\Delta u_\lambda}{u_\lambda} + \frac{\Delta(d\lambda)}{d\lambda} + \frac{4}{3}\frac{\Delta V}{V} = 0. \quad \text{... (11)}$$

Since $d\lambda$ changes in the same way as λ, we have

$$\frac{\Delta d\lambda}{d\lambda} = \frac{d\lambda}{\lambda}.$$

Substituting this in eqn. (11), we get

$$\frac{\Delta u_\lambda}{u_\lambda}+\frac{\Delta\lambda}{\lambda}+\frac{4}{3}\cdot\frac{\Delta V}{V}=0. \qquad \text{... (12)}$$

Also from eqn. (8), we have

$$\frac{1}{3}\cdot\frac{\Delta V}{V}=\frac{\Delta\lambda}{\lambda}.$$

Substituting this in (12), we get

$$\frac{\Delta u_\lambda}{u_\lambda}+\frac{\Delta\lambda}{\lambda}+\frac{4\Delta V}{\lambda}=0$$

or

$$\frac{\Delta u_\lambda}{u_\lambda}=-\frac{5\Delta\lambda}{\lambda}.$$

Integrating, we get

$$\log u_\lambda = -5\log\lambda+\log C$$

where log C is a constant of integration

or $\log u_\lambda + \log \lambda^5 = \log C$

or $\log u_\lambda \lambda^5 = \text{constant or } u_\lambda \lambda^5 = u'_\lambda \lambda'^5$ ... (13)

where u'_λ is the density of radiation of wavelength λ' (to which the radiation of wavelength λ has been transformed by the expansion) at temperature $T' = \frac{\lambda T}{\lambda'}$ (since $\lambda T = \lambda' T'$).

Obviously u_λ must be a function of T, hence the constant in eqn. *(13)* must contain T. Also we know that the constant is such that it remains constant throughout in such process of adiabatic expansion. We know from eqn. (10) that for such process λT is constant. Then clearly we may say that the constant in eqn. (13) must be a function of λT. Then eqn. (13) may be written as

$u_\lambda \lambda^5 = A' f(\lambda T)$ where A' is any constant

or

$$u_\lambda d\lambda = \frac{A'}{\lambda^5} f(\lambda T)\, d\lambda. \qquad \text{...(14)}$$

Eqn. (13) gives

$$\frac{u_\lambda}{u'_\lambda} = \frac{u'^5}{u^5} = \frac{T^5}{T'^5}. \qquad [\text{since } \lambda T = \lambda' T']$$

As emissive power $E\lambda$ is proportional to energy density, we have

$$\frac{E_\lambda}{E'_\lambda} = \frac{T^5}{T'^5}. \qquad \text{... (15)}$$

It is now apparent that if the distribution of energy at any temperature T is given, the distribution at any higher temperature can be obtained graphically by shortening each λ in the ratio

$$\frac{T}{T'}(\text{since } \lambda T = \lambda' T')$$

and increasing the corresponding E_λ in the ratio

$$\left(\frac{T}{T'}\right)^5 \left[\text{since} E_\lambda = E'_\lambda \left(\frac{T}{T'}\right)^5\right] \qquad \text{from (15)}$$

The wave interval dλ is also shortened in the ratio $\frac{T}{T'}$.

Now the curve becomes higher but more closed up such that the total area representing to the intensity is changed in the ratio $\left(\frac{T'}{T}\right)^4$.

It is found that corresponding to highest point P, i.e., point of maximum emission, in the first curve, there is a highest point in the second curve such that

$$\lambda m\ T = \lambda' m\ T' = \text{constant}.$$

which is Wien's displacement law.

PROBLEMS

1. ***Calculate the surface temperature of the sun. Given for solar radiations λm = 4753 Å and Wien's constant b = 2.898 × 10^-3^ meter-kelvin.***

Solution: We know from Wien's displacement law

$$\lambda m\ T = \text{constant} = 0.2898 \times 10^{-2} \text{ mK.} \qquad \text{... (1)}$$

For the sun given

$$\lambda m = 4753\text{Å} = 4753 \times 10^{-10} \text{ m.}$$

$$\therefore \quad \text{Temperature of sun, } T_s = \frac{0.2898 \times 10^{-2}}{4753 \times 10^{-10}} = 6097\text{K.}$$

2. *From Wein's displacement law we have*

$$\lambda_m T = \frac{hc}{5k} = b$$

(i) Calculate the constant b. What are the units of b ?

(ii) Radiation from moon gives* lm = 4700Å *and* 14 × 10^{-6} *m.

What conclusion can you draw from this ?

[h = 6.6 × 10^{-34} joule-sec., k =* 1.37 × 10^{-23} *joule/K.

c = 3 × 10^{8} *m/sec.]*

Solution: (i) As $b = \lambda m\ T$, = the unit of b is meter-kelvin (mK)

We have
$$b = \frac{hc}{5k} = \frac{6.6 \times 10^{-34} \times 3 \times 10^{8}}{5 \times 1.37 \times 10^{-23}}$$

$$= 0.002890 \text{ mK.}$$

(ii) When $\lambda m = 4700\text{Å} = 4700 \times 10^{-10}$ m, we have

$$T_1 = \frac{0.002890}{4700 \times 10^{-10}} = 6166\text{K.}$$

When
$$\lambda_m = 14 \times 10^{-6} m, T_2 = \frac{0.002890}{14 \times 10^{-4}} = 207\text{K}$$

The first one is the temperature of the sun, because sun's radiation are reflected from moon's disc and other one is moon's own temperature.

Wien's Distribution Law

We have seen from the energy-wavelength curves obtained by Lummer and Pringsheim that the energy is not uniformly distributed in the radiation spectrum of a black-body. It is found that $E\lambda\ d\lambda$, i.e. the amount of energy contained in the spectral region from wavelength λ to $\lambda + d\lambda$ emitted by a black-body at temperature T, is of the form

$$E_\lambda\, d_\lambda = \frac{A}{\lambda^5} f(\lambda T)\, d\lambda \qquad \text{... (i)}$$

where A is constant and $f(\lambda T)$ is a function of product λT.

The law may also be written as

$$E_\lambda\, d_\lambda = AT^5(\lambda T)\, d\lambda. \qquad \text{... (ii)}$$

Deduction

We have

$$E_\lambda\, d\lambda = \frac{A'}{\lambda^5} f(\lambda T)\, d\lambda \qquad \text{... (iii)}$$

But the emissive power E_λ is proportional to the energy density

i.e. $$E_\lambda\, d\lambda \approx u\lambda\, d\lambda$$

or $$E_\lambda\, d\lambda = \frac{A}{\lambda^5} f(\lambda T)\, d\lambda, \qquad \text{... (iv)}$$

which is the same as eqn. (i), i.e. Wien's' distribution law.

Eqn. (iv) may be written as

$$E_\lambda d\lambda = \frac{A}{\lambda^5} \cdot \frac{T^5}{T'^5} f(\lambda T)\, d\lambda$$

$$= AT^5(\lambda T)^{-5} f(\lambda T)\, d\lambda.$$

or $$E_\lambda\, d_\lambda = AT^5 F(\lambda T)\, d\lambda. \qquad \text{... (v)}$$

where $$F(\lambda T) = (\lambda T)^{-5} f(\lambda T).$$

Eqn. (v) is the same as eqn. (ii), i.e., second form of Wien's distribution law.

6

Radiation Pyrometers

The fact that the radiation emitted by a black-body depends solely upon its temperature, provides radiation methods for measuring high temperatures. The instruments used for measuring high temperatures by means of radiation are called radiation pyrometers. The main advantage of these pyrometers is that they are capable of measuring the tempertaure, however high it may be, and whatever be the distance of the object from the pyrometer. The instrument need not be placed in direct contact with the hot body nor it is required to be raised to the temperature of the body. These pyrometers measure the high temperature of the body without placing them in direct contact with the hot body. There are two types of radiation pyrometers

Total Radiation Pyrometers : These pyrometers measure the intensity of total radiations emitted by the body under test and then the temperature is deduced by applying Stefan's law assuming the body to be black.

Optical or Spectral Pyrometers : These pyrometers compare the intensity of radiation of a certain wavelength emitted by the body with that of the radiation of same wavelength emitted by a standard body at a known temperature. Then the temperature of the body is deduced by Wien's distribution law.

Fery's Total Radiation Pyrometer

It consists of a concave mirror M made of copper and nickled on its front surface having a central opening fitted with a telescope T. The position of the mirror can be adjusted with a rack and pinion arrangement P. D is a diaphragm placed in front of the mirror at its focus. S is a disc blackened on its side facing the mirror and one of the junctions of a thermocouple is connected to S. The other junction of the thermocouple as well as of S are protected from the direct radiations by suitable screens.

When the radiation from a hot body, whose temperature is to be determined, is incident on the mirror, the focussed image falls on the opening of D. Due to incident radiation falling on S, the galvanometer shows the deflection.

In order to obtain accurate focussing, the diaphragm is constituted by two semi-circular mirrors inclined to each other at an angle of about 5° with a central hole of about 5 mm to allow the incident radiations to pass. When the diaphragm D is viewed from eye-piece, it appears to be circular as shown in fig. (b) if the focussing is perfect. If the focussing is imperfect, the two halves appear to be displaced as shown in fig. (c). The mirror M is moved till relative displacement of the two halves disappears.

In actual use of apparatus, the adjustments are made such that the image of source formed by the mirror M is slightly greater than the hole in the diaphragm D. In this case the thermocouple measures the intensity of the image and not the total radiation and hence the readings in the galvanometer will be independent of the distance of the body from the pyrometer. It is because when the distance between the mirror and the hot body is doubled, the amount of radiation falling on the concave mirror is reduced to one-fourth according to inverse square law: but at the same time the diameter of the image is halved and its area becomes one fourth. Hence the amount of radiations received per unit area per second is constant i.e., the intensity of radiation is unaltered provided the image is larger than the aperture.

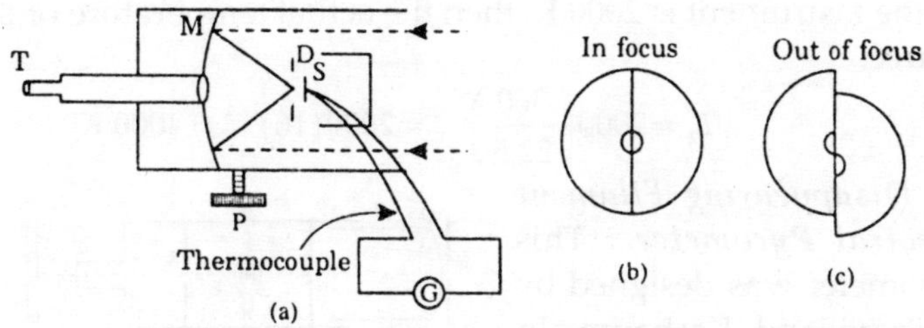

If T is the temperature of the body and T_0 that of the pyrometer case, the deflection in the galvanometer would be proportional to $\left(T^4 - T_0^4\right)$. As T_0 is smaller compared to T, the deflection in the galvanometer is proportional to T^4. In practice it is found that power lies between 3.8 and 4.2 due to stray reflections from the walls etc. Therefore, it is necessary to calibrate the instrument. This is done by using radiations from a black-body at different temperatures which are accurately measured by a thermocouple. Then a calibration curve is plotted between the logarithm of deflection θ against the logarithm of absolute temperature

$$\log \theta = K \log T.$$

The curve will be a straight line. This method gives better results.

Modification for measurement of very high temperature: In order to measure the temperatures higher than the normal range, a rotating sector is used to cut out a known fraction of radiations. An opaque disc from which a sector of q° is cut off is rotated about the axis of the pyrometer tube between the hot body and the objective. If now instrument measures the temperature T_1, the actual temperature T_2 of the source is given by calculation, in accordance of Stefan's law,

$$\frac{T_1^4}{T_2^4} = \frac{\theta^\circ}{360^\circ}$$

$$T_2 = T_1\left(\frac{360^\circ}{0^\circ}\right)^{1/4}$$

For example, if $0 = 22.5°$ and the temperature T_1 measured by the instrument is 2000 K, then the actual temperature of the source.

$$T_2 = 2000\left(\frac{360}{22.5}\right)^{1/4} = 2000(16)^{1/4} = 4000 \text{ K.}$$

Disappearing Filament Spectral Pyrometer : This pyrometer was designed by Halborn and Karbaum. In this pyrometer the intensity of the red parts of the spectrum of radiations emitted by a hot body is compared with that given out by filament of a small electric lamp. It consists of a telescope whose cross wires have been replaced by filament of an electric lamp as shown in fig. The filament is heated with the help of a battery *B*. The current passing through the filament can be adjusted by a rheostat *R*. *S, S* are two diaphragms and *O* and E are the objective and eye-piece respectively. The filament is seen through the eye-piece by placing red-glass filter between the eye and eye-piece.

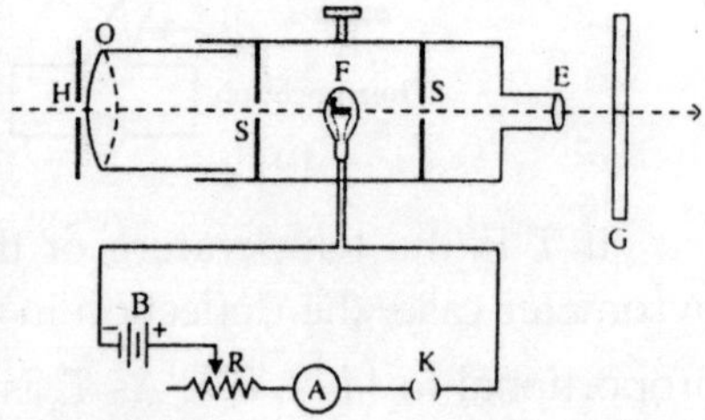

The radiations passing through the hole *H* enter the objective of telescope. The hole is focussed on the filament by adjusting the position of the telescope. Now a current is started through the filament. The filament begins to glow. The strength of the current is so adjusted that the contrast between the filament and image of the hole just disappears and the filament becomes invisible against back-ground. The current passing through the filament, when it disappears, is noted. Since both are now equally bright, they must be emitting equal amounts of heat energy per unit area per second and hence they must be at the same temperature.

Initially, the instrument is calibrated by sighting it on the hole of a constant temperature enclosure heated at various known temperatures that are measured with a standard thermocouple. The current at each stage is noted. A graph is plotted between the current and temperature. From the graph, the unknown temperature corresponding to a particular value of current passing through the filament, can be found. We can also calibrate the ammeter to measure the temperatures directly.

Solar Constant

The sun is a source of thermal energy and it emits continuously radiant energy in all directions in space, only a part of which is received by the earth. From the quantity of radiant energy received by the earth, it is possible to estimate the temperature of the sun. For the purpose it is necessary to define a constant known as Solar constant.

Solar constant is defined as the amount of solar energy received per unit time by unit area of a perfectly black surface placed at a mean distance of the earth from the sun, in the absence of the atmosphere, and at right angles to the direction of sun's rays.

Consider the sun as a black-body radiator. The temperature of the sun can be found as follows :

Let R be the mean distance of the earth from the sun and S the solar constant.

Then the total amount of radiant energy received by a sphere of radius R per second = $4\pi R^2 S$.

If r is the radius of the sun, the amount of radiant energy radiated by the sun per second = $4\pi r^2 E$ where E is the energy radiated by unit area of the sun per second.

$$4\pi R^2 S = 4\pi r^2 E$$

or $$E = \left(\frac{R}{r}\right)^2 \times S \qquad \text{... (i)}$$

If T is the temperature of the surface of the sun, then according to Stefan's law

$$E = \sigma T^4, \qquad \text{... (ii)}$$

where σ is Stefan's constant.

Comparing (i) and (ii), we get

$$\sigma T^4 = \left(\frac{R}{r}\right)^2 \times S$$

or $$T = \left[\left(\frac{R}{r}\right)^2 \times S \times \frac{1}{\sigma}\right]^{1/4} \qquad \text{... (iii)}$$

The value of solar constant is 1.4 kW/m^2 or 2 cal cm^{-1} min^{-1}

If solar constant is given in joule m^{-1} min^{-1}, then

$$T = \left[\left(\frac{R}{r}\right)^2 \times \frac{S}{60} \times \frac{1}{\sigma}\right]^{1/4} \qquad \text{... (iv)}$$

PROBLEMS

1. ***The earth receives 1.90 calories of heat per cm^2 per minute. Calculate the surface temperature of the sun. Given distance of earth from the sun = 150 × 10^6 km., solar diameter = 1.39 × 10^6 km, Stefan's constant = 1.37 × 10^{-12} cal. cm.$^{-2}$ sec.$^{-1}$ de^{-4}.***

Solution: The surface temperature of the sun is given by

$$T = \left[\left(\frac{R}{r}\right)^2 \times \frac{S}{60} \times \frac{1}{\sigma}\right]^{1/4}$$

Given $R = 150 \times 10^6$ km. $= 150 \times 10^9$ m.

$$r = \frac{1.39 \times 10^6}{2} \text{km.} = 0.695 \times 10^9 \text{ m.}$$

$$\sigma = 1.37 \times 10^{-12} \text{ cal. cm.}^{-2} \text{ sec.}^{-1} \text{ deg.}^{4}$$

$$= 1.37 \times 10^{-8} \text{ cal. cm.}^{-2} \text{ sec.}^{-1} \text{ deg.}^{4}$$

$$S = 1.90 \text{ calories}$$

$$\therefore \quad T = \left[\left(\frac{150\times10^{9}}{0.695\times10^{9}}\right)^{2}\times\frac{1.9\times10^{9}}{60}\times\frac{1}{1.37\times10^{-8}}\right]$$

$$= 5730 \text{ K.}$$

2. *The earth receives 2.0 cal. cm.$^{-2}$ min.$^{-1}$ from the sun. If angular diameter of the sun is 32′ and it is treated as a black-body, deduce its surface temperature. Given σ = 5.7 × 10^{-12} watt cm^{-2} (K)$^{-1}$.*

Solution: The temperature of the sun is given by

$$T = \left[\left(\frac{R}{r}\right)^{2}\times\frac{S}{60}\times\frac{1}{\sigma}\right]^{1/4}$$

where
$$\frac{r}{R} = \frac{\text{radius of the sun}}{\text{distance of sun from the earth}}$$

$$= \text{angular radius.}$$

Given angular diameter = 32′

$$\therefore \quad \text{Angular radius} = \frac{r}{R} = 16' = \left(\frac{16}{60}\right)^{o}$$

$$= \left(\frac{16}{60}\right)\times\frac{\pi}{180} \text{ radians.}$$

Hence
$$\frac{R}{r} = \frac{180\times60}{16\pi}$$

Also given
$$S = 2.0 \text{ cal. cm.}^{-2} \text{ min.}^{-1}$$

$$= 2\times4.2 \text{ joule} \times (10^{-2}\text{ m})^{-2} \times \text{min.}^{-1}$$

$$= 2\times4.2\times10^{4} \text{ joule m}^{-2} \text{ min.}^{-1}$$

and
$$\sigma = 5.7\times10^{-12} \text{ watt cm.}^{-2} \text{ (K)}^{-4}$$

$$= 5.7\times10^{-8} \text{ watt m}^{-2} \text{ min.}^{-1}$$

Substituting these values

$$T = \left[\left(\frac{180\times60}{16\pi}\right)^2 \times \frac{2\times4.2\times10^4}{60} \times \frac{1}{5.7\times10^{-8}}\right]^{1/4}$$

$$= 5803 \text{ K.}$$

3. *Find the pressure exerted by sunlight on earth's surface if the value of solar constant is 2 × 10⁴ cal. /min. m².*

Solution: The radiation pressure p is given by

$$p = \frac{E}{c}.$$

Here $E = 2 \times 10^4$ cal/min–m^2

$$= \frac{2\times10^4}{60} \text{cal/sec.m}^2$$

$$= \frac{2\times10^4}{60} \times 4.2 \text{ joule/sec.m}^2.$$

$$= 1.4 \times 10^3 \text{ joule/sec. m}^2.$$

Therefore, pressure exerted by sunlight on earth's surface

$$p = \frac{E}{c}$$

$$= \frac{1.4\times10^3 \times \text{joule/sec.m}^2}{3\times10^8 \text{m/s}}$$

$$= 4.66 \times 10^{-6} \text{ N/m}^2.$$

Quantum Theory of Radiation; Planck's Hypothesis

The classical theory of radiation, due to the assumption that energy change of radiators takes place continuously, could not explain the experimentally observed distribution of energy in the spectrum of a black body. Planck, therefore, was led (in 1900) to make an important hypothesis known as Planck's hypothesis. According to it, "*A black-body radiation chamber is filled up not only with radiation; but also with simple harmonic oscillators (or resonators) of the molecular dimensions, and the exchange of energy between matter and radiation does not take place continuously but discontinuously and discretely as on integral multiple of small unit of energy called the quantum or photon."* The vibration of the resonator entails one degree of freedom only. Thus, if the equipartition principle of energy were applicable, the energy associated with each resonator would have been kT units of energy which would be partly kinetic and partly potential.

Planck assumed that the energy of a photon is proportional to the frequency of radiation, i.e. $\varepsilon \propto \nu$.

or $$\varepsilon = h\nu.$$

where h *is* a universal constant called the Planck's constant, its value being equal to 6.62×10^{-34} joule sec. or 6.62×10^{-27} erg. sec. and is *dimensionally equal to the product of linear momentum and distance or the product of energy and time.*

This means that an oscillator of frequency ν can only emit or absorb the radiation in units of quanta of magnitude $h\nu$. The emission of radiation corresponds to a decrease and absorption to an increase in energy and amplitude of an oscillator, which would be partly kinetic and partly potential. Thus :

$$\varepsilon = nh\nu,\ n = 0, 1, 2, 3 \ldots\ldots \qquad \ldots (1)$$

It should be noted that energy of a photon is very small. For example, a photon of radiation of wavelength $\lambda = 10{,}000$ A contains energy equal to 2×10^{-10} joule only.

The hypothesis was originally made to explain the experimentally observed spectrum of black-body radiation. Later, it was used by Einstein to explain the photoelectric effect and the variations in specific heat of solids and gases. Bohr used it to explain atomic spectra. It has also been used to explain photo-chemical reactions.

Average Energy of Classical Oscillator in Thermal Equilibrium : If p is the momentum and x is the displacement from equilibrium position of a classical oscillator of mass m

vibrating under a force constant C, the frequency of the oscillator is

$$\nu = \frac{1}{2\pi}\sqrt{\left(\frac{C}{m}\right)} \qquad \text{... (1)}$$

and the energy of the oscillator is

$$\varepsilon = \frac{P^2}{2m} + \frac{1}{2}Cx^2. \qquad \text{... (2)}$$

If the oscillator is in thermal equilibrium with a heat reservoir at absolute temperature T, then the probability that the oscillator will have momentum and displacement in the range p to $p + dp$ and x and $x + dx$, according to Boltzmann's law, is proportional to $e^{-\varepsilon/kT}\,dp\,dx$.

$\therefore$ The average energy of classical oscillator

$$\bar{\varepsilon} = \frac{\iint \varepsilon e^{-\varepsilon/kT}\,dp\,dx}{\iint e^{-\varepsilon/kT}\,dp\,dx} \qquad \text{... (3)}$$

Using equation (2), we have

$$\bar{\varepsilon} = \frac{\iint \left(\frac{p^2}{2m} + \frac{1}{2}Cx^2\right) e^{-\left\{(p^2/2m)+\frac{1}{2}Cx^2\right\}/kT}\,dp\,dx}{\iint e^{-\left\{(p^2/2m)+\frac{1}{2}Cx^2\right\}/kT}\,dp\,dx}$$

$$= \frac{\int_{-\infty}^{\infty} \frac{p^2}{2m} e^{-p^2/2mkT}\,dp \int_{-\infty}^{+\infty} e^{-Cx^2/2kT}\,dx}{\int_{-\infty}^{+\infty} e^{-p^2/2mkT}\,dp \int_{-\infty}^{+\infty} e^{-Cx^2/2kT}\,dx}$$

$$+ \frac{\int_{-\infty}^{+\infty} \frac{1}{2}Cx^2 e^{-Cx^2/2kT}\,dx \int_{-\infty}^{+\infty} e^{-p^2/2mkT}\,dp}{\int_{-\infty}^{+\infty} e^{-Cx^2/2kT}\,dx \int_{-\infty}^{+\infty} e^{-p^2/2mkT}\,dp}$$

$$= \frac{\int_{-\infty}^{\infty} \frac{p^2}{2m} e^{-p^2/2mkT}\,dp}{\int_{-\infty}^{+\infty} e^{-p^2/2mkT}\,dp} + \frac{\int_{-\infty}^{+\infty} \frac{1}{2}Cx^2 e^{-Cx^2/2kT}\,dx}{\int_{-\infty}^{+\infty} e^{-Cx^2/2kT}\,dx}$$

$$= \frac{1}{2}kT + \frac{1}{2}kT = kT.$$

Thus the average energy of classical oscillator in thermal equilibrium at absolute temperature T is kT as given by equipartition law, since for a simple harmonic oscillator average kinetic energy is equal to the average potential energy, if the average is taken over a complete period.

Average Energy of Planck's Oscillator : If N is the total number of Planck's resonators and E their total energy, then average energy per Planck's oscillator is given by

$$\overline{\varepsilon} = \frac{E}{N}. \qquad \text{... (1)}$$

According to Maxwell's law for molecular motion if ε is a certain amount of energy, the probabilities that a system will have energies 0, ε, 2ε, ... $n\varepsilon$ ... , are in the ratio

$$e^{-0/kT} : e^{-\varepsilon/kT} : e^{-2\varepsilon/kT} .... : e^{-n\varepsilon/kT} ....etc.$$

Thus, if the number of resonators having energies 0, ε, 2ε ...... rε ...... respectively are N_0, N_1, N_2 ...... N_r ......, then from Boltzmann's canonical distribution law

$$N_0 \propto e^{-0/kT} \quad \text{or} \quad N_0 = Ce^{-0} \quad \text{or } N_0 = C$$

$$\therefore \quad N_1 \propto e^{-\varepsilon/kT} \quad \text{or} \quad N_1 = Ce^{-\varepsilon/kT} = N_0e^{-\varepsilon/kT}$$

$$N_2 \propto e^{-2\varepsilon/kT} \quad \text{or} \quad N_2 = Ce^{-2\varepsilon/kT} = N_0e^{-2\varepsilon/kT}$$

....

....

$$N_r \propto e^{-r\varepsilon/kT} \quad \text{or} \quad N_r = Ce^{-r\varepsilon/kT} = N_0e^{-r\varepsilon/kT}$$

....

....

Therefore, total number of resonators in the resonance chamber

$$N = N_0 + N_1 + N_2 + N_3 + ... + N_r + ...$$
$$= N_0 + N_0 e^{-\varepsilon/kT} + e^{-2\varepsilon/kT} + N_0 e^{-3\varepsilon/kT} + ...$$
$$+ N_0 e^{-r\varepsilon/kT} + ...$$
$$= N_0 \left[1 + e^{-\varepsilon/kT} + e^{-2\varepsilon/kT} + ... + e^{-r\varepsilon/kT} + ...\right].$$

Put $e^{-\varepsilon/kT} = y.$ ... (2)

$$\therefore \quad N = N_0 \left[1 + y + y^2 + ... + y^r + ...\right]$$
$$= \frac{N_0}{1-y} \qquad ... (3)$$

The total energy of these resonators will be

$$E = 0 \times N_0 + \varepsilon \times N_1 + 2\varepsilon \times N_2 + ... + r\varepsilon \times N_r + ...$$

or
$$E = 0 + \varepsilon \times N_0\, e^{-\varepsilon/kT} + 2\varepsilon \times N_0\, e^{-2\varepsilon/kT} + ...$$
$$= N_0\, \varepsilon\, [e^{-\varepsilon/kT} + 2e^{2t/rT} + ... + re^{-r\varepsilon/kT} + ...] \text{ [from (2)]}$$
$$= N_0\, \varepsilon y\, [1 + 2y + 3y^2 + ... + ry^{r-1} + ...]$$
$$= N_0\, \varepsilon y S. \qquad ... (4)$$

where $S = [1 + 2y + 3y^2 + ... + ry^{r-1} + ...]$ ... (5)

$\therefore \quad Sy = y + 2y^2 + ... + (r-1)\, y^{r-1} + ...$ ... (6)

Subtracting (6) from (5), we get

$$S\,(1-y) = 1 + y + y^2 + ... + y^r + ... = \frac{1}{1-y} \qquad ...(7)$$

$$\therefore \quad S = \frac{1}{(1-y)^2}. \qquad ...(8)$$

Substituting this value in equation (4), we get

$$E = \frac{N_0 \varepsilon y}{(1-y)^2}. \qquad ... (9)$$

Therefore, average energy of Planck's resonator,

$$\bar{\varepsilon} = \frac{E}{N} = \frac{N_0 \varepsilon \dfrac{y}{(1-y)^2}}{\dfrac{N_0}{1-y}}. \quad \text{from (9) and (3)}$$

$$= \frac{N_0 \varepsilon y(1-y)}{(1-y)^2 N_0}$$

$$= \frac{\varepsilon y}{1-y} = \frac{\varepsilon e^{-\varepsilon/kT}}{1-e^{-\varepsilon/kT}} \qquad \text{from (2)}$$

$$\bar{\varepsilon} = \frac{\varepsilon}{e^{\varepsilon/kT}-1} \qquad \text{... (10)}$$

Thus we see that average energy of a resonator is not kT, (as given by equipartition law), but equal to $\frac{\varepsilon}{e^{\varepsilon/kT}-1}$ according to quantum theory.

Discussion

(i) According to classical theory the energy of an oscillator may have any value, i.e., we must suppose ε to be infinitesimal. When ε tends to zero, the limiting value of $\frac{\varepsilon}{\exp(\varepsilon/kT)-1}$ is kT as given by equipartition law.

According to Planck's hypothesis of quantum theory, the value of ε is not infinitesimal, but is equal to $h\nu$; therefore, the *average energy of Planck's oscillator or resonator is* given by

$$\langle\varepsilon\rangle \text{ or } \bar{\varepsilon} = \frac{h\nu}{e^{h\nu/kT}-1} \qquad \text{... (11)}$$

(ii) If $h \to 0, \frac{h\nu}{kT} \to 0$, then equation (10) may be written as

$$\bar{\varepsilon} = \frac{h\nu}{\left[1+\frac{h\nu}{kT}+\frac{1}{2!}\left(\frac{h\nu}{kT}\right)^2+\frac{1}{3!}\left(\frac{h\nu}{kT}\right)^3+\ldots-1\right]}$$

$$= \frac{h\nu}{\frac{h\nu}{kT}\left[1+\frac{1}{2!}\cdot\frac{h\nu}{kT}+\frac{1}{3!}\left(\frac{h\nu}{kT}\right)^2+\ldots\right]}$$

$$= \frac{kT}{\left[1+\frac{1}{2!}\cdot\frac{h\nu}{kT}+\frac{1}{3!}\left(\frac{h\nu}{kT}\right)^2+\ldots\right]}$$

Obviously, if h were zero, then

$$\bar{\varepsilon} = kT$$

Thus, if h were zero, the average energy of Planck's oscillator would be the equipartition value kT.

(iii) At high temperature or low frequency $\frac{h\nu}{kT}$ is quite small

$$\therefore \quad e^{h\nu/kT} = 1 + \frac{h\nu}{kT} + \frac{1}{2!}\left(\frac{h\nu}{kT}\right)^2 + \ldots$$

$$\approx 1 + \frac{h\nu}{kT}$$

$$\therefore \quad \bar{\varepsilon} = \frac{h\nu}{1 + \frac{h\nu}{kT} - 1} = kT$$

(iv) At low temperature or high frequency $\frac{h\nu}{kT}$ is quite large

$$\therefore \quad e^{h\nu/kT} - 1 \approx e^{h\nu/kT}$$

$$\therefore \quad \bar{\varepsilon} = \frac{h\nu}{e^{h\nu/kT}} h\nu\, e^{-h\nu/kT}$$

PROBLEMS

1. ***Calculate the average energy* e *of an oscillator of frequency* 0.60 × 10^{11} *sec.*$^{-1}$ *at T* = 1800 K *treating it as (i) classical oscillator. (ii) Planck's oscillator.***

Solution :(i) The average energy of classical oscillator

$$\bar{\varepsilon} = kT$$

$$= (1.38 \times 10^{-23} \text{ joule/K}) \times 1800 \text{ K}$$

$$= 2.484 \times 10^{-20} \text{ joule.}$$

(ii) The average energy of Planck's oscillator

$$\bar{\varepsilon} = \frac{h\nu}{e^{h\nu/kT} - 1} = \frac{h\nu/kT}{e^{h\nu/kT} - 1} kT.$$

Now $$\frac{hv}{kT} = \frac{(6.6\times10^{-34})(0.60\times10^{14})}{1.38\times10^{-23}\times1800}$$

$$= \frac{3.96\times10^{-23}}{2.484\times10^{-20}} = 1.59$$

$$\therefore \quad \bar{\varepsilon} = \frac{hv}{e^{hv/kT}-1}kT = \frac{1.59}{e^{1.59}-1}\times2.484\times10^{-20}$$

$$= \frac{1.59}{(2.72)^{1.59}-1}\times2.484\times10^{-20} \text{ (since } e = 2.72)$$

$$= \frac{1.59\times2.484\times10^{-20}}{3.909} = 1.01\times10^{-20} \text{ joule.}$$

The Number of Resonators Per Unit Volume Lying in The Wavelength Range l To l + *d*l Or Frequency Range v and $v + dv$

Let us consider the radiation to be enclosed in a hollow cubic enclosure, the walls of which are perfectly reflecting. According to electromagnetic theory, the radiation is supposed to consist of a number of waves. These waves travel in all possible directions in the enclosure and undergo multiple reflections from the various walls of the enclosure. Inside the enclosure stationary vibrations are formed with the walls as nodal planes as a result of the superposition of the reflected waves with the corresponding incident waves. The formation of stationary vibrations can be understood by an analogy with the vibrations of a stretched string with both end points fixed. It is well known that the end points of the string are two nodes of stationary vibrations and only certain discrete frequencies of vibrations are allowed. If l is the length of the string, the allowed wavelengths are given by

$$\lambda = \frac{2l}{n} : n = 1, 2, 3, 4 \ldots\ldots \infty \qquad \ldots (1)$$

The corresponding allowed frequencies are

$$v = \frac{c}{\lambda} = \frac{nc}{2l}; n = 1, 2, 3, 4......\infty$$

which c is the speed of the waves.

Every allowed frequency is called a mode of vibration. The allowed modes of vibration inside the cube can be calculated just as in the case of the string. But in this case the waves are confined to a three dimensional space. We have, therefore, to think a three-dimensional picture with the walls of the cube as nodal planes. Let each side of the cube be 'a'. If we take the three intersecting edges of the cube as x, y and z axes of a cartesian coordinate system and cos α, cos β, cos γ as the direction cosines of the direction of propagation of a particular wave, then the projections of the edges of the cube on the direction of the propagation of the wave are a cos a, a cos (β and a cos γ respectively. In this case only those waves will be allowed for which all the faces of the cube form nodal planes. For this the allowed wavelength has to satisfy three conditions of the kind of eqn. (1) in which l is replaced by a cos α , a cos β and α cos γ respectively. Thus an allowed wavelength λ must satisfy.

$$\lambda = \frac{2a\cos\alpha}{n_1}, \lambda = \frac{2a\cos\beta}{n_2} \text{ and } \lambda = \frac{2a\cos\gamma}{n_3} \quad ...(2)$$

where n_1 n_2 and n_3 are positive integers.

Now, according to trigonometric condition of direction cosines

$$\cos^2\alpha + \cos^2\beta + \cos^2\gamma = 1. \quad ...(3)$$

Substituting values of cos α, cos β and cos γ from eqn. (2) in (3), we get

$$\frac{n_1^2\lambda^2}{4a^2} + \frac{n_2^2\lambda^2}{4a^2} + \frac{n_3^2\lambda^2}{4a^2} = 1$$

or $$n_1^2 + n_2^2 + n_3^2 = \frac{4a^2}{\lambda^2} = \left(\frac{2a}{\lambda}\right)^2$$

But $\lambda = \frac{c}{v}$, we have, therefore,

$$n_1^2 + n_2^2 + n_3^2 = \left(\frac{2av}{c}\right)^2 . \quad \text{... (4)}$$

This equation gives the allowed frequencies (or modes of vibration) inside the cube. According to this only those frequencies are allowed for which eqn. (4) holds with various choices of positive integral values of n_1, n_2 and n_3. Each choice of the values of n_1, n_2 and *n3* corresponds to a frequency (or mode of vibration). The total number of modes of vibration are the total number of possible sets of (n_1, n_2, n_3). The number of modes of vibration within the frequency interval v and $v + dv$ can be found with the help of equation (4).

For convenience let us first count the modes of vibration in an analogous two dimensional problem. The two dimensional analogue of eqn. (4) is

$$n_1^2 + n_2^2 = \left(\frac{2av}{c}\right)^2 . \quad \text{... (5)}$$

If we plot a graph with n_1 along x–axis and n_2 along y–axis for a given value of $\nu\left(i.e. \text{ of } \frac{2a\ \text{v}}{c}\right)$, we get the points lying on a circle since eqn. (5) represents a circle of radius $\frac{2a\ \nu}{c}$. Such a graph has been shown in fig. for $\frac{2a\ \nu}{c} = 10$. In the graph the abscissae and ordinates have been drawn corresponding to all possible integral values of n_1 and n_2 upon 10. Every intersection of those abscissa and ordinates represents an allowed set of values of n_1 and n_2 and hence a mode of vibration. The points on the circle correspond to frequency v, while those lying inside the circle correspond to frequency less

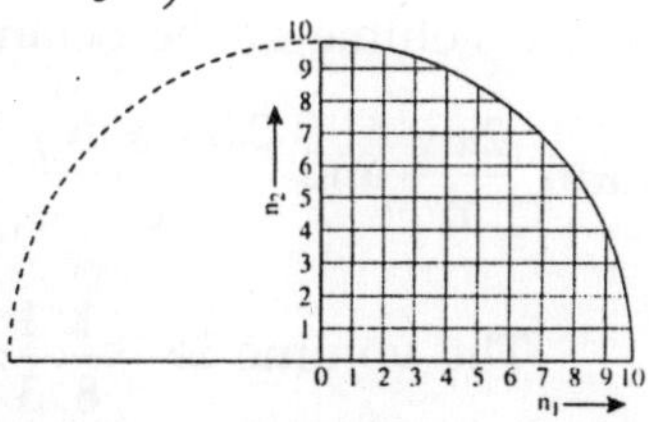

than v. As only positive values of n_1 and n_2 are allowed, we have to consider the intersections only in the positive quadrant. The lines drawn divide the quadrant in a number of unit squares. As the area of every square is unity, therefore, the number of squares is equal to the area inside the quadrant of the circle. Thus, the number of modes of vibration with frequencies less than n is equal to the area of a quadrant of a circle represented by eqn. (5). This area is $\frac{1}{4}.\pi r2 = \frac{1}{4}.\pi\left(\frac{2av}{c}\right) = \frac{\pi a^2 v^2}{c^2}$. Similarly, we can state that the number of modes of vibration with frequency interval v and $v + dv$ *is* equal to the area in the positive quadrant lying between the two circles of radii $\frac{2a\ v}{c}$ and $\frac{2a(v+dv)}{c}$.

This area is

$$\frac{1}{4}.\pi\left[\left\{\frac{2a(\text{v}+dv)}{c}\right\}^2 - \left(\frac{2av}{c}\right)^2 = \frac{1}{4}.\pi.\frac{8a^2\text{v}\,dv}{c^2} = \frac{2\pi a^2\text{v}\,dv}{c^2}\right]$$

Now considering the case of three-dimensions, we see that the circle is replaced by a sphere and a mesh of squares is replaced by a mesh of cubes. The number of modes of vibrations within frequency range v and $v + dv$ *is* now equal to the volume of the octant $\frac{1}{8}$th of a spherical shell with the radii $\frac{2a\text{ v}}{c}$ and $\frac{2a(\text{v}+dv)}{c}$.

$$\text{The volume is } = \frac{1}{8}.\frac{1}{3}.\pi\left[\left(\frac{2a(v+dv)}{c}\right)^3 - \left(\frac{2av}{c}\right)^3\right]$$

$$= \frac{1}{8}.(4.\pi)\left(\frac{2a\text{ v}}{c}\right)^2.\frac{2ad\text{ v}}{c} = \frac{4.\pi\text{ v}^2}{c^3}d\text{v}.a^3$$

$$= \frac{4\pi\text{ v}^2 d\text{v}}{c^3}V.$$

Since a^3 = V = volume of the cube.

Thus the number of modes of vibrations per unit volume within frequency range v and v+$dv = \frac{4\pi v^2 dv}{c^3}$.

Now the black body radiations travel with the velocity of light c and are transverse in character unlike sound waves in the string which are longitudinal.

In the case of transverse waves there are two possible polarisations for each wave. Therefore, the modes of vibrations of transverse waves is double as for longitudinal waves.

Therefore, for black-body radiations of electromagnetic waves, the modes of vibration within the frequency range v and v + dv

$$= \frac{2 \times 4\pi v^2 dv}{c^3}$$

$$= \frac{8\pi v^2 dv}{c^3}. \quad \text{... (6)}$$

The number of modes of vibration within wavelength range λ and λ + $d\lambda$ are

$$= \frac{8\pi}{c^3} \cdot \left(\frac{c^2}{\lambda^2}\right) \cdot \left(-\frac{c}{\lambda^2} d\lambda\right) \left(\text{Since } v = \frac{c}{\lambda} \text{ and } dv = -\frac{c}{\lambda^2}\right)$$

$$= \frac{8\pi}{\lambda^4} d\lambda. \quad \text{... (7)}$$

According to Planck, each resonator entails one degree of freedom only, therefore, the number of resonators per unit volume in the frequency range v and v + dv, from eqn. (6),

$$= \frac{8\pi v^2}{c^3} dv \quad \text{... (8)}$$

and the number of resonators per unit volume in the wavelength range λ and $\lambda + d\lambda$, from eqn. (7),

$$= \frac{8\pi}{\lambda^4} d\lambda. \quad \text{... (9)}$$

PROBLEMS

1. *Calculate the number of modes in a chamber of volume 50 c.c. in the frequency range* 4 × 10^14^ *and* 4.01 × 10^14^ *sec*^-1^.

Solution: The number of modes per c.c.

$$= \frac{8\pi v^2}{c^3} d\,\text{v}$$

$$= \frac{\left(8\times 3.14\times 4\times 10^{14}\right)^2}{\left(3\times 10^{10}\right)^3}\left(4.01\times 10^{14} - 4\times 10^{14}\right)$$

$$= \frac{8\times 3.14\times 16\times 10^{28}}{27\times 10^{30}}\times 0.10\times 10^{14}$$

$$= 1.5\times 10^{11}.$$

$\therefore$ The number of modes in the whole chamber

$$= 1.5\times 10^{11}\times 50$$

$$= 7.5\times 10^{12}.$$

2. *Calculate the number of modes of vibration in a 100 c.c. chamber in the wavelength range.*

(i) 5000 to 5002 Å (ii) 8000 to 8005 Å.

Solution: The number of modes per c.c. in the wavelength range λ to $\lambda + d\lambda$

$$= \frac{8\pi}{\lambda^4} d\lambda.$$

Therefore, the number of modes within wavelength range λ to $\lambda. + d\lambda$ in a chamber of volume V is given by

$$N_1 = \frac{8\pi V}{\lambda^4} d\lambda.$$

(i) Here $V = 100 \text{ c.c., } d\lambda = (5002 - 5000)\,\text{Å}$

$$= 2\text{Å} = 2\times 10^{-8}\text{ cm}.$$

$$\lambda = 5000\text{Å} = 5000\times 10^{-8}\text{ cm}.$$

$$N_1 = \frac{8 \times 3.14 \times 100}{(500 \times 10^{-8})^4} \times 2 \times 10^{-8}$$

$$= 3.066 \times 10^{12}.$$

(ii) Here $V = 100$ c.c., $\lambda = 8000\text{Å} = 8000 \times 10^{-8}$ cm.

and $d\lambda = (8005 - 8000)\,\text{Å} = 5\text{Å} = 5 \times 10^{-8}$ cm

$$\therefore \quad N_2 = \frac{8 \times 3.14 \times 100}{(800 \times 10^{-8})^4} \times 5 \times 10^{-8}$$

$$= 3.066 \times 10^{12}.$$

The Rayleigh-Jeans Law of Spectral Distribution of Energy : The number of modes of vibration per unit volume in the frequency range v and $v + dv$ is

$$N_2 d\text{v} = \frac{8\pi \text{v}^2}{c^3} d\text{v}. \qquad \text{... (1)}$$

Rayleigh and Jeans assumed that the law of equipartition of energy is applicable to radiation also i.e., they considered average energy of an oscillator, (i.e., per mode of vibration) as

$$\bar{\varepsilon} = kT \text{ (classical result).}$$

The energy density (i.e., energy per unit volume) within frequency range v and v + dv *is* given by

$u_V\, dv$ = Number of modes of vibration per unit volume in the frequency range v and $v + dv$ × Average energy per mode of vibration

$$u_{\bar{v}} d\text{v} = \frac{8\pi \text{v}^2 d\text{v}}{c^3} .kT. \qquad \text{... (2)}$$

This is Rayleigh-Jean's law in terms of frequency.

In terms of wavelength

$$u_\lambda d\lambda = \frac{8\pi}{c^3}.\left(\frac{c}{\lambda}\right)^2\left(\frac{-c}{\lambda^2}d\lambda\right).kT$$

$$= \frac{8\pi kT}{\lambda^4}.d\lambda. \quad \text{... (3)}$$

This is Rayleigh-Jean's law in terms of wavelength. This law explains the experimental measurement of the energy distribution at long wavelengths fairly well; but it fails for short wavelengths.

Planck's Radiation Law

Planck found an empirical formula to explain the experimentally observed distribution of energy in the spectrum of a black-body. The formula may be deduced using the following postulates

1. A black-body radiation chamber is filled up not only with radiation, but also with simple harmonic oscillators or resonators of the molecular dimensions; which cannot have any value of energy; but only energies given by

$$E = n\,hv$$

where v is the frequency of the oscillator, *h is* the Planck's constant and n is a number that can take any integral values, i.e.

$$n = 0, 1, 2, 3 \text{......}$$

2. The oscillators cannot radiate or absorb energy continuously; but an oscillator of frequency v can only radiate or absorb energy in units or quanta of magnitude *hv.* This assumption is the most revolutionary in character. In simple words this states that the exchange of energy between radiation and matter cannot take place continuously; but are limited to discrete set of values *0, hv, 2hv ...... nhv*, i.e., in multiples of some small unit, called the quantum.

The average energy of a Planck's oscillator is given by

$$\bar{\varepsilon} = \frac{h\nu}{e^{h\nu/kT} - 1} \quad \text{... (1)}$$

The number of resonators per unit volume in the frequency range ν and ν + *d*ν *is* given by

$$N = \frac{8\pi \nu^2}{c^3} d\nu \quad \text{... (2)}$$

The energy density belonging to range *dυ* can be obtained by multiplying the average energy of a Planck's oscillator by the number of resonators per unit volume, in the frequency range ν and ν + *d*ν, i.e.

$$E_\upsilon d\nu = \frac{8\pi \nu^2}{c^3} \cdot \frac{h\nu}{e^{h\upsilon/kT} - 1} d\nu$$

$$= \frac{8\pi h \nu^2}{c^3} \cdot \frac{1}{e^{h\upsilon/kT} - 1} d\nu$$

where E_ν *dν is* the energy density, i.e., total energy per unit volume belonging to the range *d*ν.

The above law was found by Planck empirically and after his name is called Planck's radiation law.

The energy density E_λ *dλ* belonging to range *dλ* can be obtained by using the relation $\upsilon = \frac{c}{\lambda}$ and hence $|d\upsilon| = \left| -\frac{c}{\lambda^2} d\lambda \right|$, *i.e.*

$$E_\lambda d\lambda = \frac{8\pi h}{c^3} \cdot \left(\frac{c}{\lambda}\right)^3 \cdot \frac{1}{e^{hc/\lambda kT} - 1} \left(-\frac{c}{\lambda^2} d\lambda\right)$$

$$= \frac{8\pi hc}{\lambda^5} \cdot \frac{1}{e^{hc/\lambda kT} - 1} d\lambda,$$

which gives the energy density for wavelength range λ and λ + *d*λ in the spectrum of a black body.

Experimental Verification of Planck's Radiation Law : The Planck's radiation law has been verified by numerous experiments. Here we shall give the experimental verification due to Lummer and Pringsheim. They used a special radiating chamber in which the conditions of "black-body radiation" or "full radiation" were practically fulfilled.

The spectrum of the black-body radiation was produced by refraction through a flour-spar prism, flour-spar being very transparent to the infrared radiations. It was necessary to apply corrections in order to convert the prismatic spectrum to the normal one. This was done by means of the known dispersion curve of flour-spar.

A concave mirror was used to obtain an image of the slit. This image was focussed on a Lummer Kurlbaum linear spectrum bolometer which was used to measure the radiant energy or distribution of energy. This bolometer was enclosed in an air-tight case in order to diminish the absorption effects due to water vapour and carbon-dioxide. The distribution of energy in the spectrum for various temperatures between 621 K and 1646 K was obtained and the curves were plotted. It was found that the distribution of energy in the spectrum of a black-body, as found by Planck's radiation law, was in perfect agreement with the experiments. Fig. represents the distribution of energy in the spectrum of black-body at 1646 K.

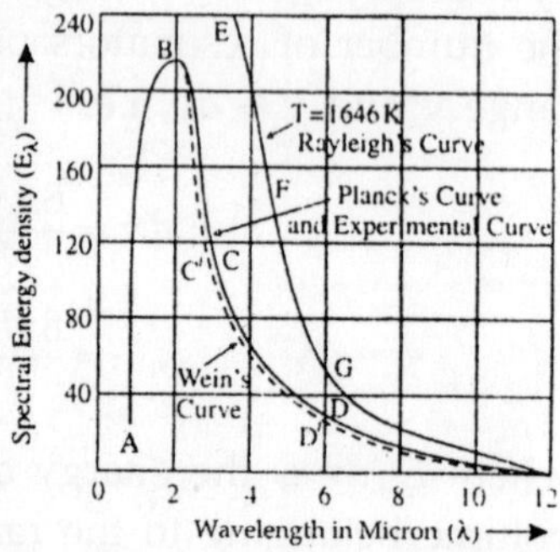

The ordinate represents the energy density per unit range of wavelength, while the abscissa represents the wavelength. In fig., the curve *ABCD* is due to experiments and Planck's radiation formula, curve *ABC'D*; is due to Wein's radiation formula and the curve *EFG is* due to Rayleigh's radiation formula. It is clear from the figure that the Planck's radiation formula is in perfect agreement with the experiments. The curve obtained by Wein's law agrees with' experimental curve only at low wavelengths while the curve obtained by Rayleigh-Jean's law agrees with experimental curve only at large wavelengths.

Wein's Law And Rayleigh–Jean's Law In Relation To Planck's Law : Planck's formula is given by

$$E_\lambda d\lambda = \frac{8\pi hc}{\lambda^5} \cdot \frac{1}{e^{hc/\lambda kT} - 1} d\lambda$$

For shorter wavelength $e^{hc/\lambda kT}$ becomes large compared to unity and hence the Planck's law reduces to

$$E_\lambda \, d\lambda = \frac{8\pi hc}{\lambda^5} \cdot \frac{1}{e^{hc/\lambda kT}} d\lambda = \frac{8\pi ch}{\lambda^5} . e^{-hc/k\lambda T} d\lambda$$

where is Wein's law.

For longer wavelengths $e^{hc/\lambda kT}$ may be approximated to $\left(1 + \frac{hc}{\lambda kT}\right)$ and hence Planck's law reduces to

$$\begin{aligned} E_\lambda \, d\lambda &= \frac{8\pi hc}{\lambda^5} \cdot \frac{1}{(1 + hc/\lambda kT - 1)} d\lambda \\ &= \frac{8\pi hc}{\lambda^5} \cdot \frac{1}{hc/\lambda kT} d\lambda \\ &= \frac{8\pi hc}{\lambda^5} \cdot \frac{\lambda kT}{hc} d\lambda \\ &= \frac{8\pi kT}{\lambda^5} d\lambda \end{aligned}$$

which is Rayleigh Jean's law.

Thus we see that Wein's law holds for shorter wavelengths while the Rayleigh Jein's law for longer wavelengths.

PROBLEMS

1. ***Deduce Wein's displacement law and Stefan's law from Planck's radiation formula.***

Solution: (i) Wein's displacement law. Planck's formula is given by

$$E_\lambda \, d\lambda = \frac{8\pi \, hc}{\lambda^5} \cdot \frac{1}{e^{hc/\lambda kT} - 1} d\lambda$$

Differentiating partially above equation with respect to wavelength λ, we get the expression of energy for unit range of wavelength, i.e..

For maximum energy of emission, i.e., for maximum value of $E\lambda$, $\frac{\partial E_\lambda}{\partial \lambda}$ must be equal to zero, i.e.,

$$\frac{-40\pi hc}{\lambda^6} \cdot \frac{1}{e^{hc/\lambda kT} - 1} + \frac{8\pi hc}{\lambda^5} \times \frac{hc}{\lambda^2 kT} \cdot \frac{e^{hc/\lambda kT}}{\left(e^{hc/\lambda kT} - 1\right)^2} = 0$$

or
$$\frac{8\pi hc}{\left(e^{hc/\lambda kT} - 1\right)\lambda^6}\left\{-5 + \frac{hc}{\lambda kT} \times \frac{e^{hc/\lambda kT}}{\left(e^{hc/\lambda kT} - 1\right)}\right\} = 0$$

or
$$-5 + \frac{hc}{\lambda kT} \times \frac{e^{hc/\lambda kT}}{\left(e^{hc/\lambda kT} - 1\right)} = 0.$$

Putting $\frac{hc}{\lambda kT} = x$, the above expression becomes

$$-5 + \frac{xe^x}{e^x - 1} = 0.$$

$$\frac{xe^x}{e^x - 1} = 5.$$

From the above equation it is clear that there must be a root in the neighbourhood of 5. Applying the method of approximation the exact value of x is found to be equal to 4.965, i.e.,

$$x = \frac{hc}{\lambda kT} = 4.965.$$

This represents the wavelength at which the energy per unit range of wavelength $\frac{\partial E_\lambda}{\partial \lambda}$ is maximum. Representing this particular wavelength by λ_m, we may write

$$\frac{hc}{\lambda_m kT} = 4.965. \quad ...(1)$$

or
$$\lambda_m . T = \frac{hc}{4.956k} = \text{constant}$$

This is Wein's displacement law.

(ii) *Stefan's Law from Planck's formula*

For it we write Planck's radiation formula in terms of frequency :

$$E_v \, d\text{v} = \frac{8\pi \text{v}^2}{c^3} . \frac{h\text{v}}{e^{hv/kT} - 1} d\text{v}$$

Therefore, the total radiant energy over all frequencies will be

$$E = \int_0^\infty E_v d\text{v} \frac{8\pi h}{c^3} \int_0^\infty \frac{v^3 d\text{v}}{e^{hv/kT} - 1}$$

Putting $\frac{h\text{v}}{kT} = x$; so that $\text{v} = \frac{kT}{h} x$ and $d\text{v} = \frac{kT}{h} dx$

Then
$$E = \frac{8\pi k^4 T^4}{c^3 h^3} \int_0^\infty \frac{x^3 dx}{e^x - 1}$$

But
$$\int_0^\infty \frac{x^3 dx}{e^x - 1} = \frac{\pi^4}{15}$$

$$\therefore \quad E = \frac{8\pi^5 k^4 T^4}{15 c^3 h^3}$$

$$= AT^4$$

where
$$A = \frac{8\pi^5 k^4}{15 c^3 h^3}$$

But
$$\frac{Ac}{4} = \sigma \rightarrow \text{Stefan's constant}$$

$$\therefore \quad E = \sigma T^4 \quad ... \text{(ii)}$$

This is Stefan's law where $\sigma = \frac{2\pi^5 k^4}{15 c^3 h^2}$ = Stefan's constant.

2. *From Planck's law deduce the value of* ν *corresponding to peak of* $E_\nu - \nu$ *curve at 1000 K. In what spectral region does this frequency lie ?*

Solution: Planck's radiation law is

$$E_\nu d\nu = \frac{8\pi h\nu^3}{c^3} \cdot \frac{1}{e^{h\nu/kT}-1} d\nu \quad \ldots (1)$$

For maximum value of E_ν, differentiating eqn. (1) with respect to ν and putting equal to zero, we get

$$\frac{\partial E}{\partial \lambda} = \frac{8\pi h}{c^3}\left[3\nu^2 \frac{1}{e^{h\nu/kT}-1} + \nu^3 \cdot \frac{\left(-\frac{h}{kT}\right) \cdot e^{h\nu/kT}}{\left(e^{h\nu/kT}-1\right)^2}\right] = 0$$

or
$$\frac{8\pi h\nu^2}{e^{h\nu/kT}-1} \cdot \left[3 - \frac{\left(\frac{h\nu}{kT}\right) . e^{h\nu/kT}}{\left(e^{h\nu/kT}-1\right)}\right] = 0$$

or
$$3 - \frac{h\nu}{kT} \cdot \frac{e^{h\nu/kT}}{e^{h\nu/kT}-1} = 0$$

Putting $\frac{h\nu}{kT} = x$, we get

$$3 - \frac{x . e^x}{e^x - 1} = 0$$

or
$$\frac{x.e^x}{e^x - 1} = 3.$$

Solving this eqn., we get approximately,

$$x = \frac{h\nu_m}{kT} = 3$$

or frequency ν_m corresponding to maximum energy of emission i.e. corresponding to peak of $E_\nu - \nu$ curve is

$$\nu_m = \frac{3kT}{h}.$$

Here T = 1000 K.

$$v_m = \frac{3\times1.38\times10^{-16}\times1000}{6.62\times10^{-27}}$$
$$= 6.25\times10^{13}\text{second}^{-1}.$$

This frequency lies in infra red spectral region.

Stefan's Constant And Wein's Constant From Planck's Law

(a) *Stefan's Constant:* The total radiant energy in unit volume of an isothermal enclosure is given by

$$E = \int_0^\infty E\lambda\, d\lambda = 8\pi\, hc \int_0^\infty \frac{d\lambda}{\lambda^5 .\left(e^{hc/\lambda kT} - 1\right)} \qquad \ldots (1)$$

(from Planck's law)

But the total radiant energy in unit volume of an isothermal enclosure is also given by

$$E = AT^4 \qquad \ldots (2)$$

where A *is* constant and T *is* the absolute temperature of the enclosure. From equations (1) and (2)

$$AT^4 = 8\pi\, hc \int_0^\infty \frac{d\lambda}{\lambda^5 \left(e^{hc/\lambda kT} - 1\right)} \qquad \ldots (3)$$

Putting $x = \frac{hc}{\lambda kT}$

$$\lambda = \frac{hc}{xkT}$$

$$\therefore \qquad d\lambda = -\frac{hcdx}{x^2kT}.$$

Substituting these values in eqn. (3), we get

$$AT^4 = 8\pi\, hc \int_0^\infty \frac{-\frac{hc\,dx}{x^2\,kT}}{\left(\frac{hc}{xkT}\right)^5 .\left(e^x - 1\right)}$$

$$= \frac{8\pi\, k^4T^4}{c^3h^3} \int_0^\infty \frac{x^3\,dx}{e^x - 1}$$

or $$A = \frac{8\pi k^4}{h^3 c^3} \int_0^\infty \frac{x^3\, dx}{e^x - 1}$$

$$= \frac{8\pi k^4 \pi^4}{15 h^3 c^3} \qquad \left[\text{Since } \int_0^\infty \frac{x^3\, dx}{e^x - 1} = \frac{\pi^4}{15}\right]$$

$$\therefore \quad A = \frac{8\pi^5 k^4}{15 h^3 c^3} \qquad \text{... (4)}$$

Since Stefan's constant (σ) is given by

$$\sigma = \frac{Ac}{4}.$$

Putting the value *of* A from eqn. (4), we have the required expression for Stefan's constant σ, given by

$$\sigma = \left(\frac{8\pi^5 k^4}{15 h^3 c^3}\right) \times \frac{c}{4}.$$

$$= \frac{2}{15} \cdot \frac{\pi^5 k^4}{h^3 c^3}.$$

(b) *Wein's Constant b :*

According to Wein's law, Wein's constant $b = \lambda_m T$

where λ_m is the wavelength for which the emitted energy is maximum.

From eqn. (1) of Ex. (17), the expression for λ_m is given by

$$\lambda_m = \frac{hc}{4.965 kT}$$

$$\therefore \quad \lambda_m T = \frac{hc}{4.965 k}$$

7

Various Aspects

For a long time it was considered that nothing could be colder than ice and hence the temperature of ice was chosen as the zero (starting point) of the scale in all earlier systems of temperature measurement. Later on, it was found that it is possible to obtain the temperature well below the ice point.

The study of the laws of perfect gases tells us that the lowest conceivable temperature on the centigrade scale is 273 degree below the melting point of ice. This temperature (–273°C) is taken as the zero of the absolute temperature scale and hence it is called *absolute zero temperature.* In this chapter, we shall discuss the principles of the methods by which the region from 0°C to absolute zero (–273°C) can be reached.

The production of very low temperatures is intimately linked with the liquefaction of gases; because all experiments in the region of very low temperatures involve the gases in the liquid or solid form. For producing very low temperatures, the usual method is to first liquify the suitable gas by purely mechanical means and then evaporating it, the latent heat of vaporisation is obtained at the expense of the heat content of the liquid itself and intense cooling results.

The following are the methods employed to produce very low temperatures:

1. *By adding salt to ice.*
2. *Cooling by evaporation of a liquid under reduced pressure.*
3. *By the adiabatic expansion of a gas.*
4. *Joule-Thomson effect.*
5. *Regenerative cooling.*
6. *By the process of adiabatic demagnetisation of a paramagnetic salt.*

The last method is useful only in the region of liquid helium temperatures and is used in producing the lowest temperatures.

Refrigerational Aspects

A refrigerator is a machine that operates in a manner opposite to that of a heat engine. In it the working substance abstracts heat from the low–temperature source (the space to be refrigerated), then some work is performed on it and finally it rejects heat to a high temperature source (the surrounding atmosphere).

Thus refrigerator is a device to transfer heat from a low temperature source to a higher temperature source with the help of some external work. The working substance which undergoes cyclic changes in the refrigerator is called the *refrigerant* and the process is known as *refrigeration.*

The schematic arrangement of a refrigerator is shown in fig. The refrigerant in the form of a liquid is stored in the liquid storage at a temperature equal to that of the condenser whose temperature is equal to that of the hot source.

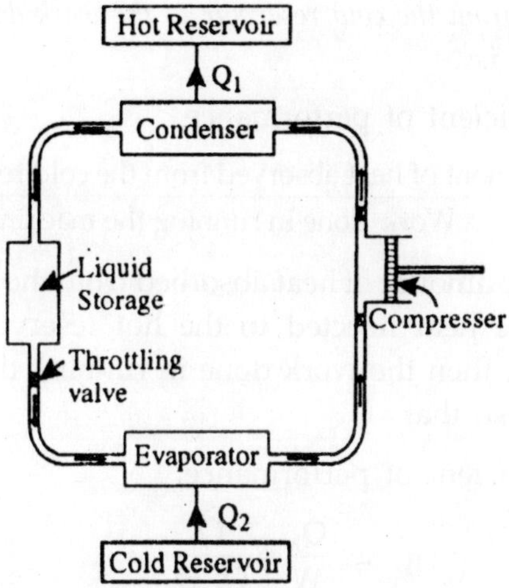

The refrigerant from the liquid storage passes through throttling valve suffering an adiabatic or Joule Thomson expansion and cools to a lower temperature. Then it passes through the evaporator and evaporates at constant temperature and pressure absorbing required latent heat from the object to be cooled. These vapours then pass through the compressor where they are compressed adiabatically by a compression pump.

The heat of compression is rejected at constant pressure by passing the vapours through the condenser maintained at low temperature, so that the working substance or refrigerant leaves the condenser as saturated or sub–cooled liquid and the fresh cycle of operations starts.

Refrigerational Performance

In a refrigerator, it is desired to extract a large amount of heat from a cold body with the expenditure of some external work. We may call, the heat extracted as output and the work as the input. It is convenient to express the efficiency of a refrigerator in terms of *coefficient of performance* which is abbreviated as *'Cop'* and *is defined as the ratio of the amount of heat absorbed from the cold reservoir to the work done in running the machinery i.e.,*

The coefficient of performance,

$$\beta = \frac{\text{Amont of heat absorbed from the cold reservation}}{\text{Work done in running the machinery}}$$

If Q_2 is the amount of heat absorbed from the cold reservoir and Q_1 is the heat rejected to the hot reservoir *during one complete cycle,* then the work done in running the machinery, $W = Q_1 - Q_2$, so that

The coefficient of performance,

$$\beta = \frac{Q_2}{W} = \frac{Q_2}{Q_1 - Q_2}.$$

But in Carnot's cycle

$$\frac{Q_1}{Q_2} = \frac{T_1}{T_2}$$

where T_2 is temperature of the cold reservoir and T_1 is the temperature of the hot reservoir.

Therefore, $$\eta = \frac{T_2}{T_1 - T_2}.$$

If η is the efficiency of a Carnot's engine operating between the same temperatures, we have

$$\eta = 1 - \frac{T_2}{T_1} = \frac{T_1 - T_2}{T_1}$$

or $$\frac{1}{\eta} = \frac{T_1}{T_1 - T_2}$$

or $$\frac{1}{\eta} - 1 = \frac{T_1}{T_1 - T_2} - 1$$

or $$\frac{1-\eta}{\eta} = \frac{T_1 - T_1 + T_2}{T_1 - T_2}$$

$$= \frac{T_2}{T_1 - T_2} = \beta$$

$\therefore$ The coefficient of performance, $\beta = \frac{1-\eta}{\eta}$.

Just as no heat engine can be more efficient than Carnot's engine, similarly the 'cop' of no refrigerator working between two reservoirs can be higher than that of Carnot's refrigerator. Coefficients of performance of actual refrigerators vary from about 2 to about 6.

PROBLEMS

1. ***A refrigerator works under a reversible cycle between the temperatures 300 K and 400 K Calculate (i) the thermal efficiency. (ii) the coefficient of performance.***

Solution: (i) Thermal efficiency

$$\eta = 1 - \frac{T_2}{T_1} = 1 - \frac{300}{400} = 0.25 \text{ or } 25\%$$

(ii) The coefficient of performance

$$= \frac{Q_2}{W} = \frac{T_2}{T_1 - T_2}$$

$$= \frac{300}{400 - 300} = 3.$$

2. ***In the above example, calculate the power for refrigerating effect of 100 Watt if the refrigerator works at (i) 100% coefficient, (ii) 60% coefficient.***

Solution: From above example

$$\beta_{max} = 3$$

(i) For 100% coefficient, $\beta_{actual} = \beta_{max} = 3$.

But $\beta = \frac{Q_2}{W}$.

Therefore the work done (power required) for refrigerating effect of 100 Watt

$$W = \frac{Q_2}{\beta_{actual}} = \frac{100}{3} = 33.3 \text{ Watt}.$$

(ii) For 60% coefficient

$$\beta_{actual} = \frac{600}{100} \times \beta_{max} = \frac{60}{100} \times 3 = 1.8$$

$$\therefore \quad W = \frac{Q_2}{\beta_{actual}} = \frac{100}{1.8} = 55.5 \text{ Watt}.$$

3. ***Calculate the least amount of work, expressed in kilowatt hours, necessary to convert 50 gm. of water at 30°C into ice at –10°C using a reservoir of cooling water at 30°C. Assume that the pressure remains constant throughout the process. Given Cp of ice. = 0.5 Kilo cal. /kg. °C and latent heat of ice = 3.4 × 10⁵ Joule. /kgm.***

Solution: T_1 = 30°C = 303 K

T_2 = – 10°C =273 – 10 = 263 K

m = 50 gm. = 0.05 kgm.

and C_p = 0.5 kilo cal./kg.°C

The quantity of heat extracted from the reservoir

$$Q_2 = 0.05 \times 30 + 0.05 \times \frac{3.4 \times 10^5}{4.2 \times 10^3} + 0.05 \times 0.5 \times 10 \text{ kilo cal.}$$

= 1.5 + 4 + 2.5 kilo cal.

= 5.75 kilo calories.

For ideal refrigerator, the coefficient of performance

$$\beta = \frac{Q_2}{W} = \frac{T_2}{T_1 - T_2}$$

Thus the required work

$$W = Q_2 \cdot \frac{T_1 - T_2}{T_2}$$

$$= 5.750 \times \frac{303 - 263}{263} = 5.75 \times \frac{40}{263} \text{ kilo calories}$$

$$= 5.750 \times \frac{40}{263} \times 4.2 \times 10^3 \text{ Joule or Watt-sec.}$$

$$(\because 1 \text{kilo cal.} = 4.2 \times 10^3 \text{ Joule})$$

$$= 5.75 \times \frac{40}{263} \times \frac{4.2 \times 10^3}{3600 \times 10^3} \text{ K.W.H.}$$

$$= 1.018 \times 10^{-3} \text{ K.W.H.}$$

Adding a Salt to Ice or Method of Freezing Mixture : This is an easy method of production of temperatures much lower than that of ice. When a salt (NaCl, say) is added to ice, it (salt) is dissolved by the water adhering to ice and more ice melts. The necessary heat for this process *viz.* that heat of solution and the latent heat necessary for melting the ice is extracted from the mixture itself.

Consequently, the temperature of the mixture falls down. This is the principle of *freezing mixtures.* This process cannot go indefinitely. The lowest temperature that can be attained by this method is fixed depending upon the salt and is called the eutectic temperature. The Eutectic temperatures for different salts are given below :

Salt	*Amount of salt in 100 gms. of mixture*	*Minimum (eutectic) temp. in °C*
$MgSO_4$	19.0 gm.	–3.9
KCl	19.7 gm.	–11.1
NH_4C1	18.6 gm.	–15.8
$NaNO_3$	37.0 gm.	–18.5
NaCl	22.0 gm.	–21.2
$CaCl_2$	29.8 gm.	–55.0

It may be observed that the temperatures thus reached are not low enough.

Cooling by Evaporation of Liquid Under Reduced Pressure : When a liquid evaporates, it requires heat for conversion from liquid to gaseous state (*i.e.*, latent heat of vaporisation).

If the vaporisation be made to take place by some contrivance and if the liquid be isolated, the latent heat of vaporisation is supplied at the expense of the total heat content of the liquid itself and hence intense cooling results. The amount of cooling depends upon the nature of liquid and the rate of evaporation.

Now-a-days the contrivance utilising the process is employed in a large number of refrigerating machines both for industrial and domestic work. To produce very low temperatures generally liquidified gases, e.g., ammonia, sulphur–dioxide etc. are used. Two types of such machines are in use

(i) vapour compression machine,

(ii) vapour absorption machine.

The former type of machines are more efficient particularly for large plants and require less initial cost, so they are more commonly used than the later type of machines. The only difference between the two types of machines consists in the manner of compressing the low pressure vapour. Here we shall describe the vapour compression machine in detail.

Vapour Compression Machine: The schematic arrangement of vapour compression machine is shown in fig. It consists of the following three principal parts:

1. The compressor G.
2. The condenser C.
3. The evaporator *E*.

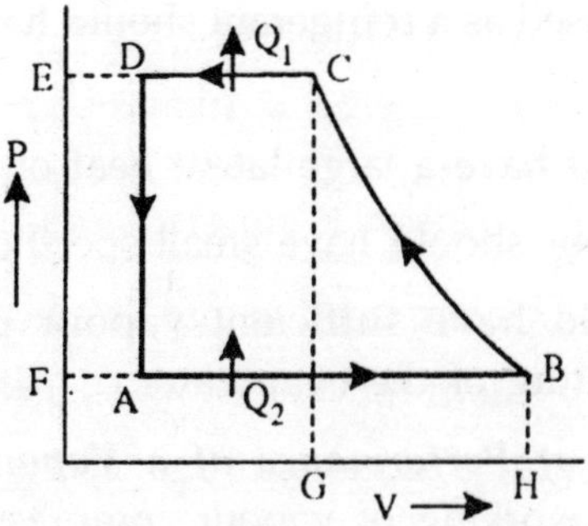

The compressor *G* is usually electrically driven and consists of a piston and two valves, one for the suction of low pressure vapour from the evaporator and the other for the discharge of the compressed vapour to the condenser.

The condenser is generally a coil of copper embedded in a vessel through which some coolant such as cold water or brine water is passing.

Working : When the piston of the compressor moves upwards, the pressure in the cylinder of compressor falls below the pressure in the evaporator. As a result the low pressure vapour is sucked through the valve and the suction pipe. When the piston moves downwards, the vapour is compressed and then delivered to the condenser C through the discharge valve and discharge pipe.

The vapour at high pressure entering the condenser maintained at low temperature liquifies in the condenser. The liquid passes through the expansion or regulator valve *V* which reduces the pressure of the liquid from the high pressure prevailing in the condenser to the low pressure in the evaporator. Due to the low pressure, the liquid evaporates extracting its latent heat either from the cold storage space or the brine water surrounding the evaporator *E*. As a result, the surrounding space or the brine water is cooled. The low pressure vapour is again sucked in by the compressor and the cycle of operations continues.

The liquid used as a refrigerant should have the following properties:

(i) It should have a large latent heat of evaporation.

(ii) Its vapour should have small specific volume.

(iii) It should have sufficient vapour pressure at the temperature of the evaporator.

Coefficient of Performance of a Vapour Compression Machine : The working of vapour compression machine is represented on a $P - V$ diagram as shown in fig. The working liquid (refrigerant like liquid NH_3, SO_2 etc.) passes through the evaporator, it evaporates to vapours at constant pressure equal to that of saturated vapours absorbing the latent heat of evaporation from the objects to be cooled. This process has been represented by the line *AB* on $P - V$ diagram. Then the vapours are compressed adiabatically to a high pressure.

This process is shown by the curve *BC*. At point *C* the pressure of the vapours is equal to that of the saturated vapour of the liquid at the temperature of the condenser. At this stage the heat of compression is removed at constant pressure by the coolant circulating in the condenser so that the vapours are condensed to liquid phase. This process is represented by the line *CD*. The liquid then flows to the storage at low pressure equal to that of the evaporator. Due to very small compressibility of the liquid, its volume remains substantially constant. This process is represented by the line *DA*. At point *A* the fresh cycle of operations starts.

The total heat content of the refrigerant at any point is given by

$$H = U + PV, \qquad \text{... (i)}$$

where U *is the internal energy.*

so that

$$dH = dU + PdV + VdP \qquad \text{... (ii)}$$

According to First Law of Thermodynamics

$$dQ = dU + PdV \qquad \text{... (iii)}$$

so that we may write

$$dH = dQ + V\,dP \qquad \text{... (iv)}$$

Now, the change from B to C is adiabatic, *i.e.*, $dQ = 0$. Therefore, the area of the curve between *BC* and the pressure axis (*i.e., ECBF*) is given by

$$\int_B^C dH = \int_B^C VdP \qquad \text{(since } dQ = 0\text{)}$$

or

$$H_C - H_B = \int_B^C VdP$$

If V_l is the constant volume of the liquid from *D* to *A* and P_2 and P_1 the pressures at points *D* and *A* respectively, the area between the line *AD* and pressure axis, (*i.e., EDAF*) is $(P_2 - P_1) V_l$. Therefore, the work done in running the machine, is given by

$$\begin{aligned} W &= \text{Area } ABCD \\ &= \text{Area } ECBF - \text{Area } EDAF \\ &= \int_B^C VdP - (P_2 - P_1)V_l \\ &= H_C - H_B - (P_2 - P_1)\,V_l \\ &\approx H_C - H_B. \end{aligned}$$

Since for a given mass the volume of the liquid is negligible in comparison to the volume of its vapours. *i.e.*,

$$H_C - H_B\ ^3\ (P_2 - P_1)\,V_l \qquad \text{... (v)}$$

The heat absorbed by the machine at low temperature is given by

$$Q_2 = \int dQ = \int (dU + PdV). \qquad \text{...(vi)}$$

But

$$dH = dU + PdV + VdP.$$

From A to B the change is isobaric, *i.e.*, $dP = 0$, so that we have

$$dH = dU + PdV \qquad \text{... (vii)}$$

Comparing (vi) and (vii),

$$Q_2 = \int_A^B dH = dH = H_B - H_A$$

Then the *Coefficient of Performance* of the vapour compression machine is given by

$$\beta = \frac{Q_2}{W} = \frac{H_B - H_A}{H_C - H_B}. \qquad \text{...(viii)}$$

PROBLEMS

1. ***In a vapour compression machine ammonia is used as a refrigerant. The temperature at which the refrigerator is maintained is –15°C; while the temperature of the coolant is 30°C. From Mollier's liquid ammonia chart the total heat contents (enthalpies) of ammonia at the points A, B and C are 77.2, 340.7 and 396.3 K cal per kg. of NH_3 respectively. Calculate the coefficient of performance of the machine and the relative efficiency of machine taking the efficiency of the ideal Carnot's refrigerator working between the same temperatures to be maximum.***

Solution: Given H_A = 77.2 K cal. per kg. of NH_3

H_B = 340.7 K cal. per kg of NH_3

H_C = 396.3 K cal. per kg. of NH_3

∴ The coefficient of performance of the machine

$$\beta = \frac{H_B - H_A}{H_C - H_B}.$$

$$= \frac{340.7 - 72.2}{396.3 - 340.7}$$

$$= \frac{263.5}{55.6}$$

$$= 4.74$$

The coefficient of performance of ideal Carnot's refrigerator is given by

$$\beta_{max} = \frac{T_2}{T_1 - T_2} = \frac{(273-15)}{(273+30)-(273-15)} = \frac{258}{45} = 5.73$$

$\therefore$ The relative efficiency of the machine $= \frac{4.75}{5.73} = 0.827$

$= 82.7\%$

Process of Cooling

During an adiabatic process the gas is perfectly insulated from surroundings *i.e.*, the gas can neither give heat nor take heat from the surroundings. So, when the gas expands adiabatically against some external pressure, it has to do work at the expense of its internal energy. As the internal energy of a gas is a function of its temperature, it is obvious that during adiabatic expansion of the gas its temperature must fall. If a highly compressed gas is suddenly allowed to expand adiabatically, it has to do a large amount of external work and so the cooling effect will be large.

If P is the pressure and T the temperature of a perfect gas, then during adiabatic process

$$PV^{\gamma} = \text{constant}$$

γ being the ratio of two specific heats.

Using gas equation $PV = RT$, we must have

$$TP^{(1-\gamma)/\gamma} = \text{constant.}$$

If T_1 and P_1 are the initial temperature and pressure of a given mass of the gas, while T_2 and P_2 are its final temperature and pressure, then we have

$$T_1 P_1^{(1-\gamma)/\gamma} = T_2 P_2^{(1-\gamma)/\gamma}$$

or the final temperature $T_2 = T_1\left(\frac{P_1}{P_2}\right)^{(1-\gamma)/\gamma}$ ... (i)

or $$T_2 = T_2 = T_1\left(\frac{P_2}{P_1}\right)^{(\gamma-1)/\gamma}$$ ...(ii)

From above expression we see that at ordinary initial temperature T_1 and for high pressure difference, the expanding mass of the gas may be liquified or even solidified. For example, if CO_2 gas at room temperature compressed to a pressure of 150 atmosphere is suddenly released to atmospheric pressure, the final temperature is

$$T_2 = T_1\left(\frac{P_2}{P_1}\right)^{(\gamma-1)/\gamma} = 300\left[\frac{1}{150}\right]^{(1.3-1)/1.8} = 96\,\text{K}.$$

This temperature is so low, that the expanding mass of CO_2 gets solidified, which is known as dry ice.

Process of Expansion

The machine utilizing the above principle is called gas expansion machine. The working of the machine is represented in fig.

It consists of the following four main parts

1. Compressor *P* driven by an electric motor.
2. Water cooler *C*.
3. An expansion engine *E*.
4. A reservoir or cold storage space *R*.

Working : The gas or air is first compressed in the compressor P adiabatically. The compressed gas is passed through a water cooler C where it is cooled to a suitable temperature at constant pressure. This cooled gas is then passed to expansion engine where it suffers adiabatic expansion to the pressure equal to that of the intake of the compressor and is cooled to low temperature. This cold gas is further passed to the reservoir or the cold storage space where it absorbs heat at constant pressure and finally returns to the compressor so that a fresh cycle of operation starts. The power for the compressor is partly obtained from expansion engine and partly from the electric motor.

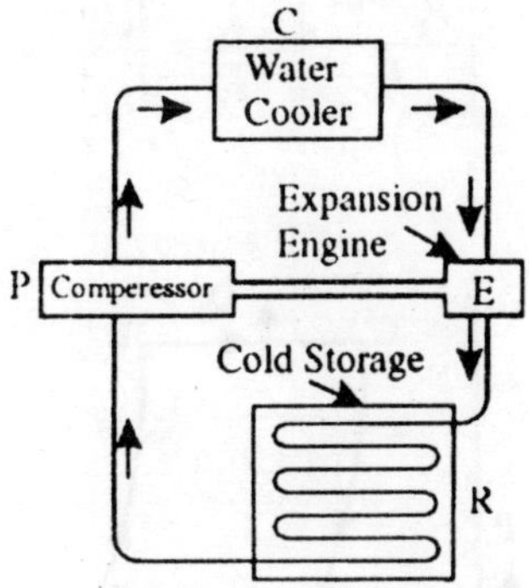

Coefficient of Performance of a Gas Expansion Machine: The working cycle of the machine is represented on a $P - V$ diagram in Fig. The gas at low pressure P_1 (generally atmospheric pressure) enters the compressor at point A where the temperature of the gas is T_α. Here the gas is compressed adiabatically to pressure P_2, high compared to P_1, temperature of the gas being T_b (point B). The compressed gas is cooled at this pressure through the cooler to the temperature T_c, of the coolant (point C).

The gas is now passed to expansion engine where it is allowed to expand adiabatically along path CD to the point D where the pressure and temperature of the gas are P_1 and T_d respectively. This cooled gas is now passed through the reservoir R from which it absorbs heat at constant pressure along path DA. As the gas reaches the state A, the cycle is completed and a fresh one starts.

Let the quantity of the working gas be 1 gm. molecule.

The quantity of heat given to the coolant.

$$Q_1 = CP\,(T_b - T_c) \qquad \text{.... (iii)}$$

The quantity of heat absorbed by the gas,

$$Q_2 = C_P\,(T_a - T_d). \qquad \text{... (iv)}$$

From equation (ii) we have

$$\left[\frac{P_2}{P_1}\right]^{(\gamma-1)/\gamma} = \frac{T_c}{T_d} = \frac{T_b}{T_a} \qquad \text{... (v)}$$

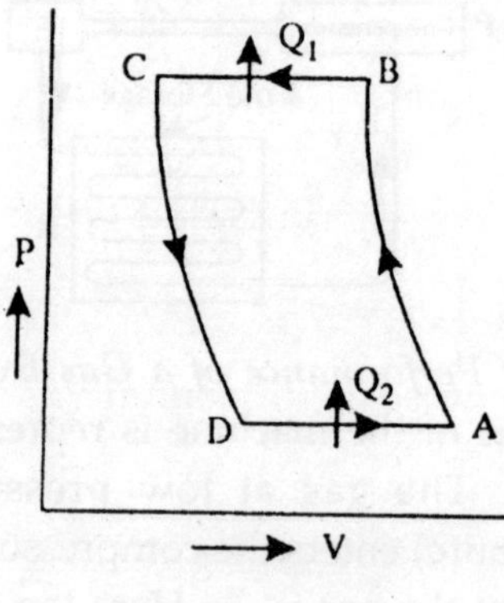

The coefficient of performance of the machine is given by

$$\beta = \frac{Q_2}{W} = \frac{Q_2}{Q_1 - Q_2}$$

$$= \frac{C_P(T_a - T_d)}{(T_b - T_c) - (T_a - T_d)} = \frac{1}{\dfrac{(T_b - T_c)}{(T_a - T_d)} - 1}$$

or

$$\beta = \frac{1}{\left[\left\{\dfrac{P_2}{P_1}\right\}^{(\gamma-1)/\gamma} - 1\right]}.$$

PROBLEMS

1. ***Calculate the coefficient of performance of an adiabatic expansion refrigerator using oxygen at compression ratio 5. γ for oxygen =1.40 and $(5)^{2/7}$ =1.58.***

Solution: Coefficient of performance of a gas expansion machine

$$= \frac{1}{\left[\left\{\frac{P_2}{P_1}\right\}^{(\gamma-1)/\gamma} - 1\right]}.$$

Given that $\frac{P_2}{P_1} = 5$ and $\gamma = 1.4$

$$\therefore \qquad \beta = \frac{1}{5^{\left(\frac{1.4-1}{1.4}\right)}-1} = \frac{1}{5^{\frac{0.4}{1.4}}-1}$$

$$= \frac{1}{(5)^{2/7}-1} = \frac{1}{1.58-1} = \frac{1}{0.58}$$

$$= 1.72.$$

2. ***The compressor of an air expansion machine intakes air at 1 atmosphere and releases at 5 atmosphere pressure. Calculate the coefficient of performance of the machine. If the power necessary to drive the machine is 1.5 H.P., find the amount of heat removed per minute. y for air = 1.4.***

Solution: The coefficient of performance is given by

$$\beta = \frac{Q_2}{W} = \frac{1}{\left(\frac{P_2}{P_1}\right)^{\frac{(\gamma-1)}{\gamma}} - 1}$$

Here, P_2 = 5 atoms., P_1 = 1 atoms.

$\gamma = 1.4$

and W = 1.5 H.P = 1.5 × 746 Joule/sec.

$$\beta = \frac{1}{\left(\frac{5}{1}\right)^{\frac{(1.4-1)}{1.4}} - 1} = \frac{1}{\left(\frac{5}{1}\right)^{0.4/1.4} - 1} = 1.72$$

Heat removed per minute

$$Q_2 = \beta W$$

$$= 1.72 \times 1.5 \times 746 \times 60 \text{ Joule–min}^{-1}$$

$$= \frac{115 \times 10^8}{4.2} \text{Cal./min.}$$

$$= 27.3 \text{ kilo calorie minute.}$$

Joule–Thomson Expansion : When a gas is allowed to escape adiabatically, through a' porous plug, from the region of constant high pressure to the region at constant low pressure, it undergoes a change of temperature. This phenomenon is called *Joule–Thomson effect* and the process is called the *Joule–Thomson* or *adiabatic throttling*. The change in temperature depends upon the nature of the gas and the initial temperature. This effect has been discussed in detail in chapter 1. Here we shall summarize the results of the experiment

1. On suffering Joule–Thomson expansion, all gases undergo a change in temperature. At ordinary temperatures most of the gases show a cooling effect; while hydrogen and helium show heating effect. However, *at sufficiently low temperature all gases show a cooling effect on suffering Joule–Thomson expansion.*
2. The fall in temperature is directly proportional to the difference of pressure on the two sides of the porous plug.
3. The fall in temperature per atmosphere difference of pressure decreases as the initial temperature of the gas increases.
4. The fall in temperature becomes zero at a particular temperature, called the *temperature of inversion, i.e., at*

a temperature where the gases show neither cooling nor heating effect. *At temperature above the temperature of inversion, the gases show heating effect.* The temperature of inversion is different for different gases. The important conclusion that can be drawn from above results is :

Any gas below its temperature of inversion will cool on suffering Joule–Thomson expansion.

From chapter 1, the fall in temperature dT for a gas obeying Vander–Waal's equation of state is given by

$$dT = \frac{1}{Cp}\left(\frac{2\alpha}{RT} - b\right)dP \qquad \text{...(i)}$$

where dP is the difference in pressure on the two sides of the porous plug.

From above expression, it is clear that

(i) dT will be positive if $\frac{2\alpha}{RT} - b$ is positive *i.e.* $\frac{2\alpha}{RT} > b$

or $T < \frac{2\alpha}{Rb}$.

Therefore, cooling will take place if the temperature of the gas is less than $2a/Rb$.

(ii) dT *will* be zero if $\frac{2\alpha}{RT} - b = 0$, *i.e.* $T = \frac{2\alpha}{Rb}$.

Therefore, the fall in temperature will be zero if the temperature of the gas is equal to $2a/Rb$. This is called the *temperature of inversion* and is represented by T_i. Thus, the temperature of inversion,

$$T_i = \left(\frac{2a}{Rb}\right). \qquad \text{...(ii)}_i$$

(iii) dT will be negative if $\left(\frac{2\alpha}{RT} - b\right)$ *is* negative

i.e., $b > \frac{2\alpha}{RT}$ or $T > \frac{2\alpha}{Rb}$ or $T > T_i$.

Therefore, heating will take place if the temperature of the gas is greater than the temperature of inversion.

Inversion Curve : Let us suppose that a series of Joule–Thomson experiments is performed on the same gas in the following manner. The initial pressure P_1 and temperature T_1 (on the high pressure side of the plug) are chosen arbitrarily and the pressure P_2 on the other side of the plug is fixed at any desired value less than P_1 and then temperature T_2 of the escaping gas is measured. Keeping P_1 and T_1 the same, P_2 is varied to another value and corresponding T_2 is measured. This is done for a number of times for different values of P_2 and the different values of the final temperature T_2 in each case measured.

If now a graph is plotted between various values of P_2 and the corresponding values of T_2, a curve is obtained as shown in fig. One point on the curve is (P_1, T_1) and the others, being indicated by *a, b, c, d ...*, represent various P_2 's and T_2's. Since in the Joule–Thomson expansion, the initial and final enthalpies are equal, the enthalpy at all points *a, b, c* is the same and the curve drawn through these points thus represents a constant enthalpy (isenthalpic curve). An isenthalpic curve is the locus of all points representing equilibrium state of the same enthalpy.

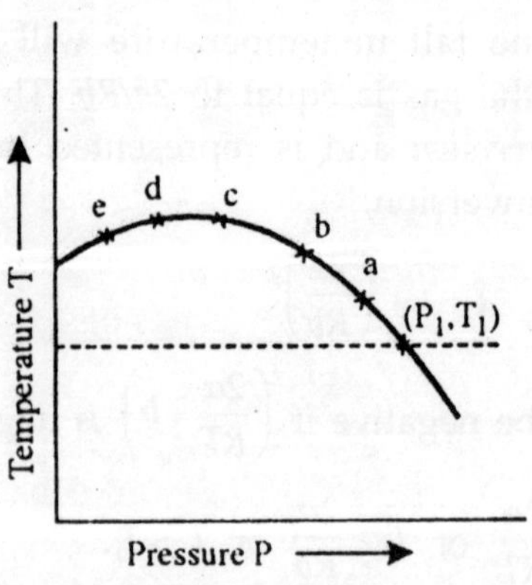

Now the temperature T_1 on the high pressure side of the plug is changed to another value keeping P_2 the same. P_2 is again varied and corresponding T_2's measured. By performing other series of experiments with the initial temperature T_1 same for one series; but different for different series, a family of curves corresponding to different values of enthalpy H are obtained as shown in fig. If the temperature is not too great, such curves pass through a maximum called the *inversion point.* The locus of all the inversion points is called the *inversion curve* (shown in fig.).

Since towards the left of the maximum of either curve of constant enthalpy (isenthalpic curve) the gas shows cooling effect, the region inside inversion curve is the *cooling region,* for in this region gas shows *cooling effect.* On the other hand, the region outside the inversion curve is called the *region of heating* because in this region the gas shows heating effect. It can be seen that at any given pressure less than the extreme value, there are two inversion temperatures, One is the "ordinary" inversion temperature (higher value) at which the heating effect gives place to cooling; and the other a much lower temperature at which cooling gives place again to heating.

The ratio of these two inversion temperatures for a Vander Waal's gas is approximately 9. In order to produce cooling effect, the gas must be precooled below a temperature where the inversion curve cuts the temperature axis on $P-T$ diagram, *i.e.,* the maximum inversion temperature. For most of the gases ordinary temperature is already below the maximum inversion temperature, and so precooling is not necessary in these cases. However, for hydrogen and helium, even the maximum inversion temperatures are much below the ordinary temperature and hence at ordinary temperatures they show a heating effect. These gases must, therefore, be precooled upto –80°C and –243°C respectively and then they will also show a cooling effect.

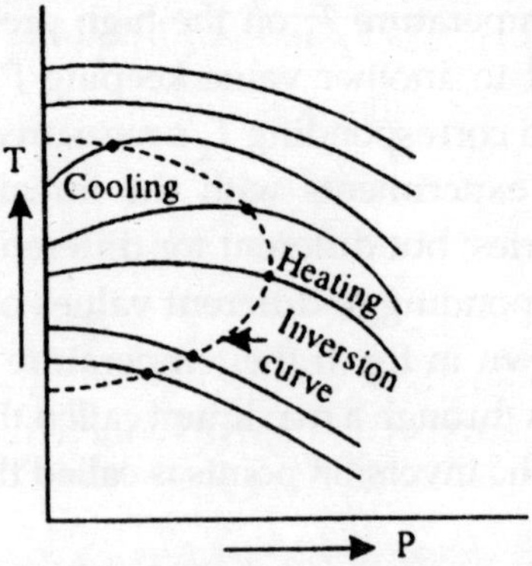

It must be carefully noted that the inversion temperature is not constant for a given gas but varies with the initial pressure of the gas.

PROBLEMS

1. *For one mole of hydrogen, the Vander Waal's constants are*

$$a = 0.245 \frac{litre^2 \times atoms}{mole^2}$$

and $b = 2.67 \times 10^{-2}$ *litre* $mole^{-1}$.

Calculate its temperature of inversion. Given that R = 2 cal. $mole^{-1}$ K^{-1} *and 1 atmosphere pressure =* 10^6 *dyne/*cm^2.

Solution: The temperature of inversion T_i is given by

$$T_i = \frac{2\alpha}{Rb}.$$

Here $a = 0.245$ litre^2 atoms/mole^2

$= 0.245 \times 10^6 \times 10^6$ cm^4 dynes/mole^2

$= 0.245 \times 10^{12}$ cm^4 dyne/mole^2

$b = 2.67 \times 10^{-2}$ litre/mole

$= 26.7$ cm^3/mole

and $R = 2$ cal/mole K

$= 2 \times 4.2 \times 10^7$ erg/mole K

Hence $$T_i = \frac{2 \times 0.245 \times 10^{12}}{2 \times 4.2 \times 10^7 \times 26.7}$$

$$= 220 \text{ K.}$$

Regenerative Cooling : The cooling produced in Joule–Thomson expansion of a gas depends on the difference of pressure on the two sides of the porous plug and the initial temperature. For most of the gases the Joule–Thomson cooling is vary small. For example, at temperature of 20°C when the pressure on the two sides is 50 atmosphere and 1 atmosphere respectively, the air cools only by 11.7°C.

In order to increase the cooling, the pressure difference was increased; but the cooling did not increase very large. For example, at a temperature of 20°C when the pressure on the two sides was kept 210 atmosphere and 1 atmosphere, the air cooled only by 42°C. However, the cooling effect can be intensified by employing the process called the *regenerative cooling*.

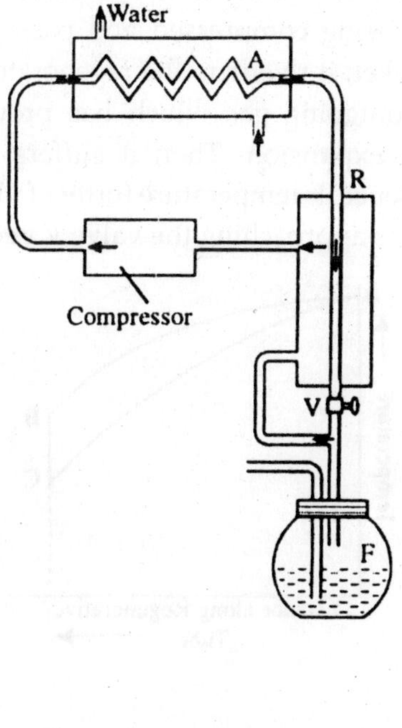

In this process a portion of the gas which suffers Joule–Thomson expansion and becomes cooled is employed to cool other portions of the incoming gas before the latter reaches the nozzle to suffer Joule–Thomson expansion. After suffering the Joule–Thomson expansion at the nozzle, the incoming gas becomes further cooled. By a continuous application of this process, the temperature of the gas coming out of the nozzle, falls continuously till the gas starts liquifying on the low pressure side. The commutative process used to cool a gas continuously is called the *regenerative cooling*.

The schematic arrangement of regenerative principle is shown in fig. The gas is compressed to a high pressure in the compressor. The heat of compression is removed by passing the gas in a water cooled jacket A. Then this gas enters the inner tube of regenerator *R* and suffers Joule–Thomson expansion at the valve *V* so that it cools by a small amount. This cooled gas returns by the outer tube and absorbs heat from the incoming high pressure gas before it reaches the compressor almost at the same temperature as the incoming gas at *R*. After being compressed and passing through the water cooled jacket, it re-enters the regenerator and is further cooled by the outgoing gas which has previously suffered Joule–Thomson expansion. Then it suffers Joule–Thomson expansion and hence its temperature further falls. As the process continues, the gas approaching the valve *V* becomes more and

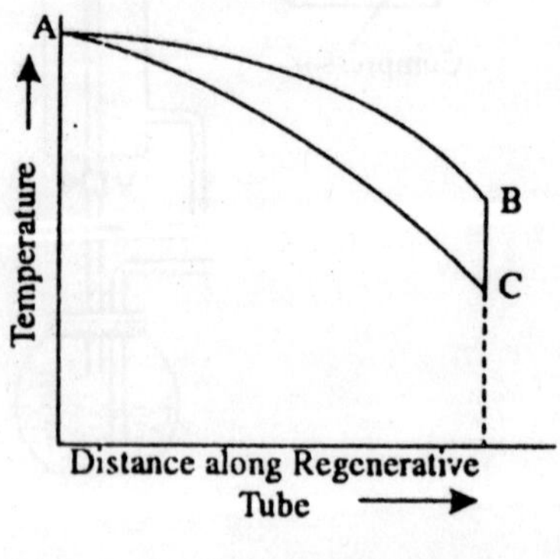

more cooled till the Joule–Thomson cooling at V is sufficient to liquify the gas. A Portion of the escaping gas is condensed in the flask F. At this stage the temperature of the whole apparatus becomes steady and may be represented by the curve shown in fig.

The part AB represents the continuous decrease of the temperature of the gas as we move from R to V through the inner tube; while part BC represents the Joule–Thomson cooling, B being the temperature of the gas before suffering Joule–Thomson expansion and C that of the gas after suffering Joule–Thomson *expansion.* The part CA represents the temperature of the low pressure gas which is less than the temperature of the adjacent high pressure gas at every point of the tube. Thus, the low pressure gas extracts heat from the incoming gas. The principle of regenerative cooling was used by Hampson and Linde. Linde used a number of coaxial tubes and Hampson used spirals instead of a single tube. The gas coming out of each tube suffers Joule–Thomson expansion, thus cooling the incoming gas.

Efficiency of the Regenerative Machines : The air at temperature T_e and pressure P_e enters the valve V and expands to pressure P_L and cools to temperature T_L. A certain fraction x is liquified while the remaining fraction $(1- x)$ goes through the outer tube at pressure P_f and temperature T_f. In the steady state $P_L = P_f$ and if the heat exchanger is efficient, the temperature of the gas, as it leaves, is only slightly different from the temperature at which it entered *i.e.*, $T_e \approx T_f$

Now since, the enthalpy on the two sides of the throttle valve is equal; the enthalpy of the entering gas must be equal to the enthalpy of x gm. of emerging liquid plus the enthalpy of $1- x$ gm. of emerging gas. Thus, if

H_e = Enthalpy of entering gas at T_e, P_e.

H_L = Enthalpy of emerging liquid at T_L, P_L.

H_f = Enthalpy of emerging gas at T_f

Then $H_e = xH_L + (1 - x) H_f$

or
$$x = \frac{H_f - H_e}{H_f - H_L}.$$

Now the state '*L*' is already known since it corresponds to the boiling point of the liquid at 1 atmospheric pressure. If the state '*f*' is supposed to be prescribed, the only variable quantity is the state '*e*'. So x will be maximum when H_e is minimum.

PROBLEMS

1. *The enthalpies of a certain gas before and after suffering Joule Thomson expansion through a throttle valve are 77.2 Gals. and 106.8 Gals. respectively. Deduce the fraction of the gas liquified if the enthalpy of emerging liquid be 55.4 calories.*

Solution: The fraction of the gas liquified in a regenerative machine is given by

$$x = \frac{H_f - H_e}{H_f - H_L}.$$

$$H_e = 77.2 \text{ cal.}, H_f = 106.8 \text{ cal}$$

and $H_L = 55.4$ calories

$$\therefore \quad x = \frac{106.8 - 77.2}{106.8 - 55.4}$$

$$= \frac{29.6}{51.5}.$$

$$= 0.57$$

Adiabatic Demagnetisation of A Paramagnetic Salt : In 1926, Debye and Giauque showed theoretically that the temperatures considerably below 1 K could be obtained by the process known as *adiabatic demagnetisation of a paramagnetic salt* (*i.e.*, those substances for which the magnetic susceptibility X is small; but positive).

Principle : The process of magnetising a substance involves doing work on the substance in aligning the elementary magnets in the direction of the external field. This work is stored in the substance in the form of magnetic energy. If the magnetised substance is demagnetised adiabatically, it has to do work and the required energy for it is drawn from within itself. As a result, the substance cools. This phenomenon (drop in temperature due to adiabatic demagnetisation) is called 'magneto–caloric effect'. Considerable cooling can be produced by employing strong magnetic field and low initial temperature. This follows from Curie law, according to which the magnetic susceptibility of a substance is inversely proportional to the absolute temperature, *i.e.*, $\chi \propto - 1/T$ or $\chi = C/T$ where C is a constant called curie constant. The final temperature is determined by measuring the magnetic susceptibility of the substance and then calculating from Curie Law.

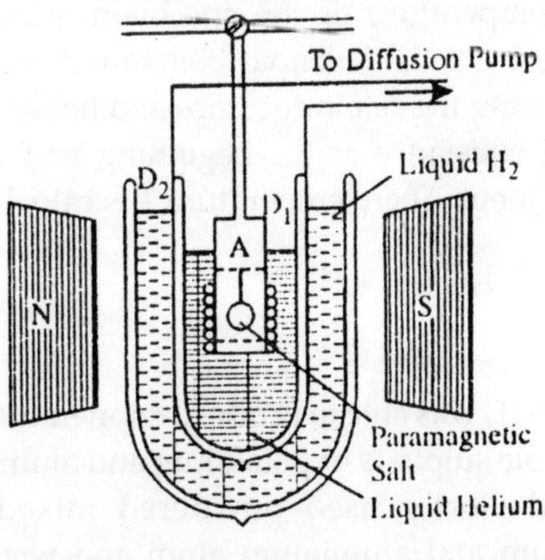

Apparatus : Debye, Giauque and Mcdougall were able to produce the temperatures below 1 K with the help of paramagnetic salt gadolinium sulphate [$Gd_2(SO_4)_3.8H_2O$]. The experimental arrangement is shown in fig. The sample (paramagnetic salt, gadolinium sulphate, say) is suspended in a vessel surrounded by liquid helium contained in a Dewar flask D_1 which, in turn, is surrounded by a Dewar flask D_2

containing liquid hydrogen. The whole arrangement is placed between the poles of a strong electromagnet.

Procedure

(i) The magnetic field is switched on, so that the specimen is magnetised.

(ii) The heat due to magnetisation is removed by first introducing hydrogen gas into *A* and then pumping it off with a high vacuum pump, so that the specimen is thermally isolated. In the mean while the specimen picks up the temperature of the liquid helium *i.e.*, the specimen and the liquid helium are at the same temperature.

(iii) Then the magnetic field is switched off. Adiabatic demagnetisation of the specimen takes place and its temperature falls.

(iv) The temperature of the specimen is determined by fitting a coaxial solenoid coil round the tube *A* and measuring the self inductance and hence susceptibility of the substance at the beginning and at the end of experiment. Then temperature *T* is calculated by Curie law.

$$\chi = \frac{C}{T}$$

Hass, in 1941, was able to produce temperatures upto 0.002 *K* using a double sulphate of potassium and aluminium. Klerk, Stenland and Gorter used powdered mixed crystals of chromium alum and aluminium alum and went down to a temperature of 0.0014 K

Theory : When a paramagnetic solid is placed in a magnetising field *H*, its elementary magnetic dipoles get aligned in the direction of the field. The magnetic moment per unit volume thus produced is called the intensity of magnetisation and is denoted by *I*. According to Curie's Law, this intensity of magnetisation is directly proportional to magnetising field H and inversely proportional to temperature T of the paramagnetic solid, *i.e.*,

$$I \propto \frac{H}{T}$$

or

$$I = C\left(\frac{H}{T}\right)$$

where C is a constant, called Curie constant.

If V is the volume of 1 mole (molar volume), then intensity of magnetisation for 1 mole of paramagnetic substance

$$M = IV = CV\left(\frac{H}{T}\right) \qquad ...(1)$$

Let us now suppose that 1 mole of paramagnetic solid is placed in magnetising field H. Then its thermodynamic behaviour can be expressed in terms of thermodynamical variables P, V, T and S. But in thermodynamic system, an increase in pressure P results in a decrease in volume V whereas here an increase in H results in an increase in M. Therefore, replacing P by $-H$ and V by M in Maxwell's third thermodynamic relation

$$\left(\frac{\partial T}{\partial P}\right)_S = \left(\frac{\partial V}{\partial S}\right)_P$$

We get

$$\left(\frac{\partial T}{\partial H}\right)_S = -\left(\frac{\partial M}{\partial S}\right)_H$$

or

$$\left(\frac{\partial T}{\partial P}\right)_S = \frac{-(\partial M/\partial T)_H}{(\partial S/\partial T)_H} = \frac{-T\left(\frac{\partial M}{\partial T}\right)_H}{T\left(\frac{\partial S}{\partial T}\right)_H}$$

$$= -T\left(\frac{\partial M}{\partial T}\right)_H \Big/ \left(\frac{\partial Q}{\partial T}\right)_H = -\frac{T}{C_H}\left(\frac{\partial M}{\partial T}\right)_H$$

where $C_H = \left(\frac{\partial Q}{\partial T}\right)_H$ Joule/mole K is the specific heat of the substance under constant field.

For an adiabatic change in the field (entropy S constant), we may write

$$dT = -\frac{T}{C_H}\left(\frac{\partial M}{\partial T}\right)_H dH$$

For a finite change of field from H_1 to H_2, we get the temperature change

$$\Delta T = -\frac{T}{C_H}\int_{H_1}^{H_2}\left(\frac{\partial M}{\partial T}\right)_H dH \qquad \text{...(ii)}$$

Putting the differential of eq. (1)

$$\left(\frac{\partial M}{\partial T}\right)_H = -\frac{CVH}{T^2}$$

in eq. (ii), we get

$$\Delta T = -\frac{T}{C_H}\left(-\frac{CV}{T^2}\right)\int_{H_1}^{H_2} H.dH$$

$$= \frac{CV}{C_H.T}\int_{H_1}^{H_2} H.dH$$

$$= \frac{CV}{2C_H T}\left(H_2^2 - H_1^2\right) \qquad \text{...(iii)}$$

If the magnetic field is reduced from H to zero, *i.e.*, $H_1 = H$ and $H_2 = 0$, then change in temperature is given by

$$\boxed{\Delta T = -\frac{CV}{2C_H T}.H^2} \qquad \text{...(iv)}$$

Following conclusions can be drawn from eq. (iv)

(i) The temperature of paramagnetic solid decreases as the field is reduced (indicated by negative sign of ΔT.)

(ii) Greater is the initial field H and lower the initial temperature T, greater is the temperature drop ΔT.

It is to be noted that here CV is the Curie constant per mole. If 1 gm of paramagnetic substance is taken, then CV would stand for Curie constant per gm.

T–S Diagram : In fig. we have illustrated the cooling produced by adiabatic demagnetisation using a temperature entropy diagram. Here the entropy of the substance is represented schematically as a function of temperature for different external fields $H = 0$ and $H = H$. The change in entropy during isothermal magnetisation is represented by the arrow AB. This shows that during magnetisation the entropy of the substances decreases. Now, if the field is switched off, the substance demagnetises adiabatially, so that the temperature of the system falls while entropy remains constant as represented by arrow BC.

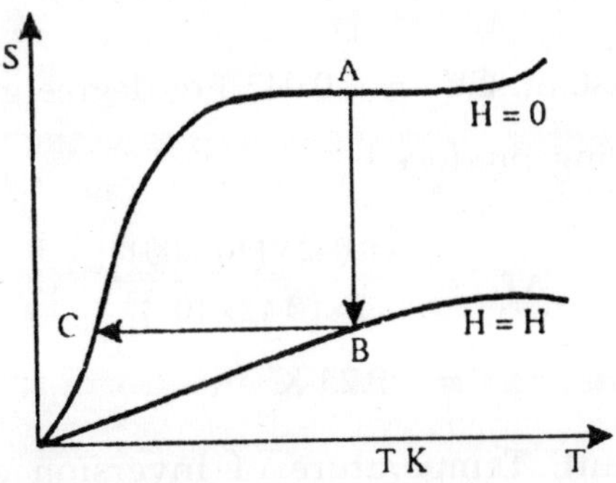

With the help of this process, temperature of the order of 10^{-4} can be reached. The measurement of temperature in this region is a matter of great difficulty but what is actually done is to assume the validity of Curie's law and calculate from the measurement of susceptibility of the sample a curie temperature $T^* = \frac{C}{\chi}$ (which is also called magnetic temperature). Later from Curie temperature, the thermodynamical temperature T is determined.

Temperatures of the order of 10^{-6} K may be obtained by making use of nuclear magnetism. However, absolute zero can never be attained.

PROBLEMS

1. ***Calculate the cooling produced by adiabatic demagnetisation of a paramagnetic salt as the field is reduced from 10000 Oersted to zero at initial temperature of 2K. Given Curie constant per gm. mole per c.c. = 0.042 Erg degree/gm oersted² and C_H = 0.42 joule gm⁻¹ deg⁻¹.***

Solution: The final temperature is given by

$$\Delta T = -\frac{CV}{2C_H T}.H^2$$

Given $T_1 = 2\text{K}, C_H = 0.42 \text{ joule gm}^{-1} \text{deg}^{-1}$

$$= 0.42 \times 10^7 \text{ erg gm}^{-1} \text{deg}^{-1}$$

and

Curie constant, CV = 0.042 Erg degree gm^{-1} Oersted^{-2}

Thus cooling produced

$$\Delta T = -\frac{0.042\times(10{,}000)^2}{2\times(0.42\times10^7)^2}$$

$$= -0.25 \text{ K}$$

Boyle Point, Temperature of Inversion and Critical Temperature of a Gas

Boyle Temperature : According to Boyle and Charles the equation of state for a perfect gas is

$$PV = RT \qquad \text{... (i)}$$

where R is gas constant for 1 gm. mole.

According to this relation, for a given quantity of gas and given temperature, the value of the product PV determined over a wide range of pressure should be constant (Boyle's law). But it was found experimentally by Amegat that for air, oxygen, nitrogen and carbon–dioxide, the value of the product *PV* decreases initially with increasing pressure, becomes a minimum at a particular pressure and then increases with increase in pressure, but for hydrogen the product *PV* increases steadily with increase in pressure, as shown in fig.

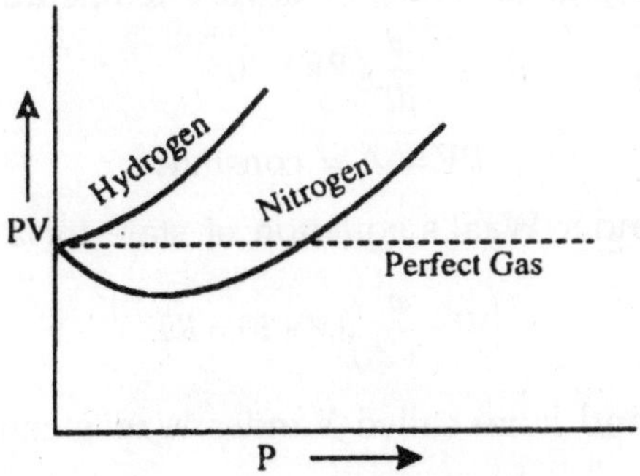

Thus, the Boyle's law is obeyed only under ideal conditions, *i.e.*, at high temperature and low pressure.

Attempts were made to modify Boyle's law. K. Onnes, after investigating the behaviour of various gases, suggested that the behaviour of all actual gases on compression at constant temperature can be represented by an equation of the type

$$PV = A + BP + CP^2 + DP^3 + \dots \qquad \dots \text{(ii)}$$

where *A, B, C, D*, etc. are constant for a given temperature depending on the nature of the gas and are called *Virial coefficients*. *A* is called the first Virial coefficient, *B* the second Virial coefficient and *C* the third and so on. These coefficients are in decreasing order of magnitude *i.e.*, $A > B > C > D$. The ratio B/A is of the order of 10^{-3} while C/A is of the order 10^{-5}. *A* is simply equal to *RT* because as $P \to 0$, equation (ii) reduces to equation (i). The second, third and fourth terms become negligible in comparison to first at very low pressure and then gases obey Boyle's law accurately–the lower the pressure the higher is the degree of accuracy.

The coefficient B has a special importance. For all gases it varies in a similar manner. Its value is negative at low temperatures, but gradually increases to zero and becomes positive as the temperature is increased. The temperature at which coefficient B is zero is called the *Boyle temperature*, represented by T_B; because at this temperature Boyle's law holds over a wide range of pressures provided the constants C, D... are negligible. Thus, at Boyle's temperature

$$B = \frac{d}{dP}(PV) = 0$$

and $$PV = A = \text{constant}.$$

Now, Vander Waal's equation of state for a real gas is

$$\left(P + \frac{a}{V^2}\right)(V - b) = RT \qquad \text{...(iii)}$$

where a and b are called Vander Waal's constants.

$$PV - Pb + \frac{a}{V} - \frac{ab}{V^2} = RT$$

or in the form of equation (ii)

or $$PV = RT = Pb - \frac{\alpha}{V} + \frac{\alpha b}{V^2} \qquad \text{...(iv)}$$

We can write approximately

$$V = \frac{RT}{P}. \qquad \text{... (v)}$$

Putting this value of V in equation (iv), we get

$$PV = RT + Pb - \frac{a}{RT}P + \frac{ab}{R^2T^2}P^2.$$

or $$PV = RT + \left(b - \frac{a}{RT}\right)P + \frac{ab}{R^2T^2}P^2. \qquad \text{...(vi)}$$

Comparing this with equation (ii) we get

$$A = RT, B = b - \frac{\alpha}{RT} \text{ and } C = \frac{\alpha b}{R^2T^2} \qquad \text{...(vii)}$$

At Boyle temperature $T = T_B$ and $B = 0$, therefore, we have

$$= b - \frac{a}{RT_B}$$

or $$T_B = \frac{a}{Rb} \quad \text{...(viii)}$$

This equation gives the value of Boyle's temperature for all gases obeying Vander Waal's equation of state. Clearly Boyle temperature depends upon a and b which are different for different gases. Hence, *Boyle temperature is different for different gases.* For nitrogen its value is 50°C, for hydrogen it is –164°C while for helium it is –250°C.

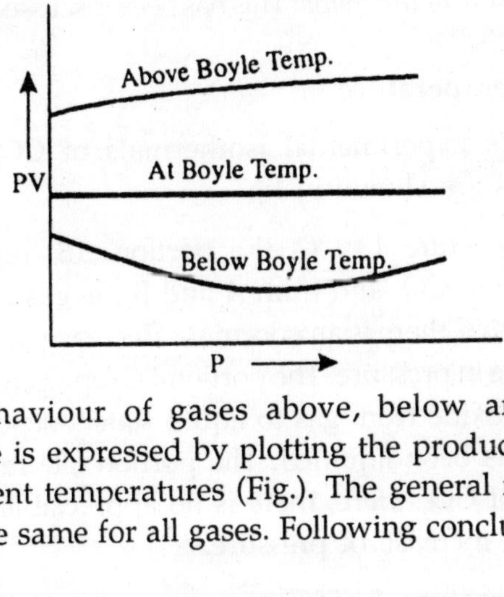

The behaviour of gases above, below and at Boyle temperature is expressed by plotting the product PV against P for different temperatures (Fig.). The general nature of the curves is the same for all gases. Following conclusions can be drawn:

(i) *Below Boyle Temperature:* The value of PV first decreases (as B is negative) with increase in P, reaches a minimum at a particular pressure and then begins to increase. It is on account of this fact that at high pressures the terms CP^2, DP^3 etc. of eq. (ii) become more effective and are positive and thus PV increases with increase in P after reaching a minimum. It suggests the existence of intermolecular attraction because below Boyle temperature the gases are highly compressible.

(ii) *Above Boyle Temperature* : The product PV increases with increase in P from the very beginning. Coefficient B is positive above Boyle temperature and thus the terms on the right hand side of eq. (ii) are positive. As a result value of PV increases with increase in P. It indicates that intermolecular attraction is now less significant.

(iii) *At Boyle Temperature* : The product PV is practically constant for a long range of pressure but at very high pressure it begins to increase. Thus, Boyle's law is obeyed over a wide range of pressures.

Temperature of Inversion: This has been discussed in article 4.6.

Types of Temperature

Andrew's experimental isothermals of CO_2 at various temperatures are shown in fig.

At temperature 13.1°C, the portion *AB*⁻ represents the gaseous state of CO_2 and from *A* and B the gas obeys Boyle's law. From *B* to C there is an enormous decrease in volume with slight increase in pressure. The portion *BC* represents the change of Carbon dioxide from gas to liquid state. At point C whole of the gas has been liquified. The portion *CD* represents the liquid state of CO_2. Here, there is no appreciable decrease in volume with increase of pressure.

At temperature 21.5°C, the curve is similar, but the horizontal portion *FG* of the curve is less than that at 13.1°C. At temperature 31.1°C the horizontal portion of the curve vanishes indicating that the gas cannot be liquified any more and its volume decreases with increase in pressure.

The curves at temperature 32.5°C and 35.5°C have the same behaviour; but the portion of the large compressibility decreases. At temperature 48.1°C the curve is smooth and similar to permanent gas obeying Boyle's law.

From fig. it is clear that CO_2 liquifies only at a temperature below 31.1°C. This temperature is called *critical temperature.* This is different for different gases and may be defined as follows :

The temperature below which a gas can be liquified by the application of pressure alone is called the critical temperature.

Above this temperature the gas cannot be liquified howsoever high the applied pressure may be.

In fig. the pressure, volume and temperature corresponding to the point *P* are called the critical pressure (P_C), the critical volume (V_C) and the critical temperature (T_C) respectively.

The tangent at the critical point *P* is horizontal so that *dP/dV* at critical point will be zero. This point *P* is called the point of inflexion. According to Vander–Waal's equation of state

$$\left(P+\frac{a}{V^2}\right)(V-b)=RT \qquad \text{...(ix)}$$

$$P = \frac{RT}{(V-b)}-\frac{a}{V^2} \qquad \text{...(x)}$$

Differentiating *P* with respect to *V*, we get

$$\frac{dP}{dV}=-\frac{RT}{(V-b)^2}-\frac{2a}{V^3} \qquad \text{...(xi)}$$

At critical point

$$\frac{dP}{dV} = 0, T = T_C \text{ and } V = V_C$$

$\therefore$ Equation (xi) gives

$$\frac{-RT_C}{(V_C-b)^2}+\frac{2\alpha}{V_C^3}=0$$

or $$\frac{2a}{V_C^3} = \frac{-RT_C}{(V_C-b)^2} \qquad \text{... (xii)}$$

Differentiating again equation (xi) w.r. to *V*, we get

$$\frac{d^2P}{dV^2} = \frac{2RT}{(V-b)^2} - \frac{6a}{V^4} \qquad \text{... (xiii)}$$

At critical point $\frac{d^2P}{dV^2} = 0, T = T_C, V = V_C$, therefore equation (xiii) gives

$$\frac{2RT_C}{(V_C-b)^3} - \frac{6\alpha}{V_C^4} = 0$$

or

$$\frac{6a}{V_C^4} = \frac{2RT_C}{(V_C-b)^3} \qquad \text{... (xiv)}$$

Dividing (xii) by (xiv), we get

$$\frac{V_C}{3} = \frac{V_C - b}{2}$$

or

$$V_C = 3b$$

Substituting the value of VC in equation (xii), we get

$$\frac{2a}{27b^3} = \frac{RT_C}{(3b-b)^2}$$

or

$$T_C = \frac{8a}{27Rb} \qquad \text{...(xvi)}$$

Substituting VC and TC from (xv) and (xvi) in (x), we get

$$P_C = \frac{R \times 8a}{27RB(2b)} - \frac{a}{9b^2}$$

$$P_C = \frac{a}{27b^2} \qquad \text{...(xvii)}$$

Equations (xv), (xvi) and (xvii) give the expression for critical volume, critical temperature and critical pressure respectively.

Relation Between Boyle Temperature, Temperature of Inversion and Critical Temperature : From equation (ii) of article 4.6 the temperature of inversion

$$T_i = \frac{2a}{Rb}. \quad \text{...(i)}$$

From equations (viii) and (xvi) of article, the Boyle temperature and critical temperature are given by

$$T_B = \frac{a}{Rb} \quad \text{...(ii)}$$

and $$T_C = \frac{8a}{27Rb} \quad \text{...(iii)}$$

From (i) and (ii)

$$T_i = 2T_B \quad \text{...(iv)}$$

From (i) and (iii)

$$\frac{T_i}{T_C} = \frac{\frac{2a}{Rb}}{\frac{8a}{27Rb}} = \frac{27}{4}$$

$$= 6.75$$

$$T_i = 6.75\ T_C. \quad \text{... (v)}$$

Thus, the temperature of inversion is very much higher than the critical temperature. The experimental value of T_i / T_C for actual gases is just less than 6. For example, in case of hydrogen, T_i = 190 K while T_C = 33 K only.

Equations (iv) and (v) give

$$T_i = 2T_B = 6.75\ T_C. \quad \text{... (vi)}$$

This relation indicates that

$$T_i > TB > TC.$$

Reduced Equation of State-Law of Corresponding States : Two gases are said to be in *corresponding states* if they have the same values of the ratios of their actual pressure, volume and temperature and critical pressure, critical volume and critical temperature. It means that

$$\frac{P_1}{P_{C_1}} = \frac{P_2}{P_{C_2}}, \frac{V_1}{V_{C_1}} = \frac{V_2}{V_{C_2}} \text{ and } \frac{T_1}{T_{C_1}} = \frac{T_2}{T_{C_2}}.$$

If we express pressure, volume and temperature as fractions of their respective values, an equation of state can be obtained which is the same for all the gases. The principle is known as law of *corresponding states* and the equation of state so obtained is called the *reduced equation of state.* Thus putting

$$\frac{P}{P_C} = \alpha, \quad \frac{V}{V_C} = \beta, \quad \frac{T}{T_C} = \gamma,$$

i.e.,

$$P = \alpha P_{C'} \; V = OV_{C'} \; T = \gamma T_C$$

where α, R and γ are mere numbers (*i.e.,* fractions) and are known as *reduced variables of state.* We get from Vander Waal's equation

$$\left(P + \frac{a}{V^2}\right)(V - b) = RT$$

$$\left(\alpha P_C + \frac{\alpha}{\beta^2 V_C^2}\right)(\beta V_C - b) = R_\gamma T_C$$

But

$$P_C = \frac{a}{27b^2}, V_C = 3b \text{ and } T_C = \frac{8a}{27Rb}$$

$\therefore$

$$\left[\frac{\alpha a}{27b^2} + \frac{a}{\beta^2 .9b^2}(\beta.3b - b)\right] = \frac{R.\gamma 8.a}{27Rb}$$

or

$$\frac{\alpha}{27} + \frac{1}{9\beta^2}(3\beta - 1) = \frac{8\gamma}{27}$$

or

$$\left(\alpha + \frac{3}{\beta^2}\right)(3\beta - 1) = 8\gamma$$

This is the reduced equation of state and is true for all gases because it does not contain the Vander Waal's constants a and b which vary from gas to gas. *When two gases have the same* *values for any two of the three reduced variables (α , β and γ) they also have the same value for the third variable and are said to be in corresponding states.* This is the statement of the *law of corresponding states.*

PROBLEMS

1. *Calculate the critical temperature and Boyle temperature of CO_2, for which the Vander Waal's constant are given to be*

$$\alpha = 0.0072 \text{ and } b = 0.002$$

The unit of pressure is atmosphere and the unit of volume is that of a gm. mol. of the gas at NTP.

Solution: From Vander Waal's eq. we have

$$\left(P + \frac{a}{V^2}\right)(V - b) = RT.$$

At N.T.P, $P = 1$ and $V = 1$ (given)

$$\therefore \quad \left[1 + \frac{.0072}{1^2}\right](1 - .002) - R \times 273$$

This gives,

$$R = \frac{1.0072 \times .998}{273}$$

The critical temperature is given

$$T_C = \frac{8a}{27Rb}$$

$$= \frac{8 \times 0.0072}{27 \times \left(\frac{1.0072 \times .998}{273}\right) \times 0.002}$$

$$= 289.6\ K.$$

$$= (289.6 - 273)°C = 16.6°C.$$

The Boyle temperature T_B in terms of critical temperature, from eq. (vi) is given by

$$T_B = \frac{6.75}{2} T_C = \frac{6.25}{2} \times 289.6\,\text{K}$$

$$= 977.4\ \text{K}$$

$$= (977.4 - 27.3)°\text{C}$$

$$= 664.4°\text{C}$$

2. ***Show that for a gas obeying Vander Waal's equation***

$$\frac{RT_C}{P_C V_C} = \frac{8}{3}$$

where the symbols have their usual meanings.

Solution: From equations (xv), (xvi) and (xvii) of article, we have

$$V_C = 3b$$

$$T_C = \frac{8a}{27Rb}$$

and

$$P_C = \frac{a}{27b^2}$$

$$\therefore \quad \frac{RT_C}{P_C V_C} = \frac{R \times \left(\frac{8a}{27Rb}\right)}{\left(\frac{a}{27b^2}\right)(3b)}$$

$$= \frac{8}{3}.$$

3. ***Calculate the critical temperature of a gas for which the Vander Waal's constants are***

$$a = 0.00874,\ b = 0.0023.$$

and the gas constant is given by

$$273\ R = 1.00646.$$

Solution: The critical temperature is given by

$$T_C = \frac{8a}{27Rb}$$

given $a = 0.00874, b = 0.0023$ ...(i)

and $R = \frac{1.00646}{273}$.

Substituting these values in equation (1), we get

$$T_C = \frac{8 \times 0.00874}{27 \times \left(\frac{1.00646}{273}\right) \times 0.0023}$$

$$= 305.5°C.$$

Process of Liquefaction

The systematic work on the liquefaction of gases started in 1823 when Faraday could succeed in liquefying chlorine, carbon dioxide, hydrogen sulphide etc. by cooling them below the room temperature and applying pressure. But the gases such as oxygen, nitrogen, hydrogen; and helium etc. could not be liquefied even by applying very high pressure and hence were termed as the "permanent gases".

Andrew's experiments on CO_2 in 1862 showed that a gas must be cooled below critical temperature before being liquefied under pressure. It resulted in the liquefaction of oxygen in 1877 by Pictet. But the real progress in the field was obtained by coupling Joule–Thomson effect and regenerative cooling. Thus, using it, Linde could liquefy oxygen in 1896 on the commercial basis. Hydrogen and helium were also liquefied on the same principle.

We shall discuss the liquefaction of these gases one by one.

Liquefaction of Oxygen: Cascade Process Kamerlingh Onnes arranged a cascade to liquefy oxygen as shown in fig.

The apparatus consists of three compression pumps P_1, P_2 and P_3 and three compression chambers *A*, *B* and *C*. The compression chambers are joined in series. The pump P_1

compresses methyl chloride gas which is cooled by cold water circulating in the outer jacket of *A*. As the critical temperature of methyl chloride is 143°C, it is liquefied at room temperature. This liquid methyl chloride passes through valve V_l into the outer jacket of *B* which is connected to the suction side of P_1. As a result of reduced pressure, the liquid methyl chloride evaporates and the temperature in *B* reaches –90°C.

The inner tube of *B* is filled with ethylene gas which is now cooled to –90°C. As the critical temperature of ethylene is 10°C, it gets liquified during the compression stroke of pump P_2 and passes through volume V_2 into the outer jacket of *C*. There, it is allowed to evaporate under reduced pressure with the help of pump P_2 resulting in the fall of temperature to –160°C. This is below the critical temperature of oxygen (–118°C) which is filled in the inner tube of *C*. Oxygen, therefore, gets liquefied during the compression stroke of pump P_3 and is collected in the Dewar flask *D*.

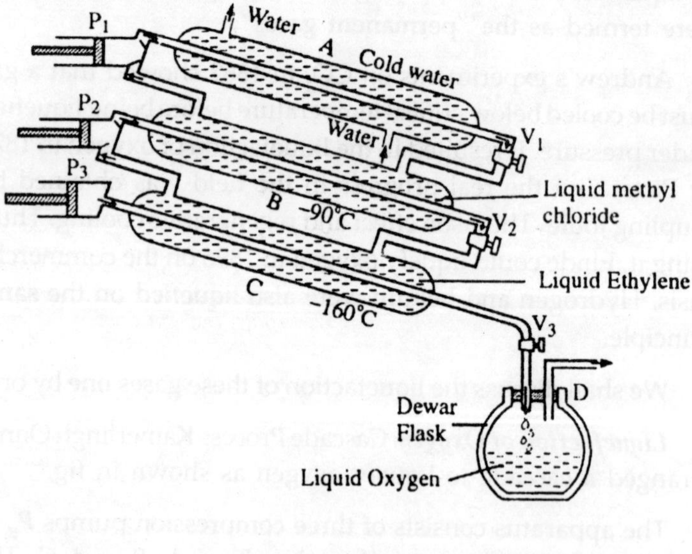

Liquefaction of Air–Linde's Process: Linde liquefied air in 1896 using Joule–Thomson effect. The schematic arrangement of his process is shown in fig. First of all air is compressed to a pressure of about 25 atmosphere by pump P_1 and is then passed through a tube surrounded by a jacket through which cold water is circulated. This compressed air is passed through KOH solution so that CO_2 and water vapour may be absorbed and removed from air.

Then the air, coming out of KOH solution, free from CO_2 and water vapour, is compressed to a pressure of 200 atmosphere by the pump P_2 and is passed through a spiral tube surrounded by a jacket containing a freezing mixture, so that air at high pressure is cooled to –20°C. This cooled air at high pressure is allowed to come out of the nozzle N_1, so that Joule–Thomson effect takes place and the incoming air is cooled. This cooled air is circulated back to pump P_2 where it is further compressed and again it is passed through the nozzle N_1 and is further cooled. This cooled air is further passed through the nozzle N_2 (from high pressure to low pressure) so that it is further cooled.

As the process continues, and after coming out of the nozzle N_2, it gets liquefied and is collected in Dewar flask D. The unliquified air is again circulated back to the pump P_1 and the process is repeated again and again.

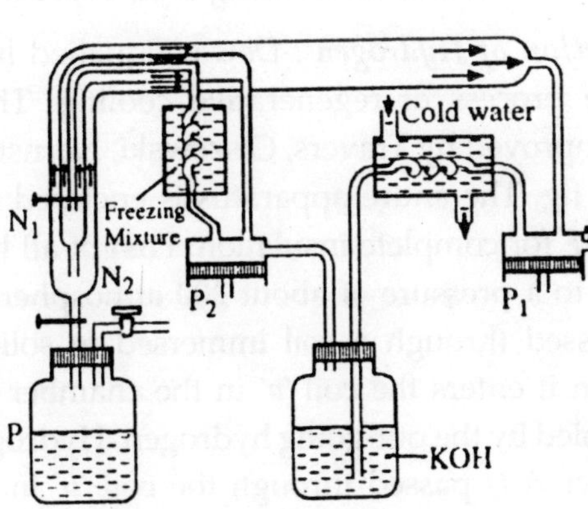

The entire arrangement is packed in cotton wool to avoid any conduction or radiation.

Uses of Liquid Air

(i) Liquid air is used to cool the substances to very low temperatures.

(ii) It is used in chemical industries for separation and purification of gases.

(iii) It is also used in developing high vacuum by solidifying a gas less volatile than air.

Constituents of liquid air (oxygen, nitrogen, argon, neon etc.) can be easily separated and are used commercially as follows

(i) Oxygen is employed in metal cutting such as breaking up of old ships and other machines. It is also used in breathing apparatus in aeroplanes, mines and diving suits.

(ii) Nitrogen is used in the manufacture of ammonia and fertilizers.

(iii) Argon is used in gas filled electric lamps.

(iv) Neon is used in discharge tubes required for illumination and advertising sign boards.

Liquefaction of Hydrogen : Dewar liquefied hydrogen in 1898 by the process of regenerative cooling. The Dewar's apparatus improved by Travers, Olszewski, Nernst and others is shown in fig. The entire apparatus is enclosed in an outer Dewar flask *F* for complete insulation. First of all hydrogen is compressed to a pressure of about 200 atmosphere by pump *P* and is passed through a coil immersed in solid CO_2 and alcohol. Then it enters the coil *'a'* in the chamber *A* where it is further cooled by the out-going hydrogen. Hydrogen coming from chamber *A* is passed through the coil *'b'* in chamber *B* containing liquid air so that hydrogen is further cooled. This hydrogen (coming from chamber B) is passed through coil *'c'* in the chamber *C* where liquid air is allowed to boil under reduced pressure (10 cm. of Hg) so that in this chamber hydrogen is cooled to a temperature of –200°C. This cooled hydrogen passes through the regenerative coil *D* and nozzle *N* so that hydrogen is cooled further due to Joule–Thomson effect. The cooled hydrogen coming from the nozzle *N* is allowed to circulate back to the pump *P*. The process of regenerative cooling continues and after some time hydrogen gets liquefied and is collected in the Dewar flask *D*.

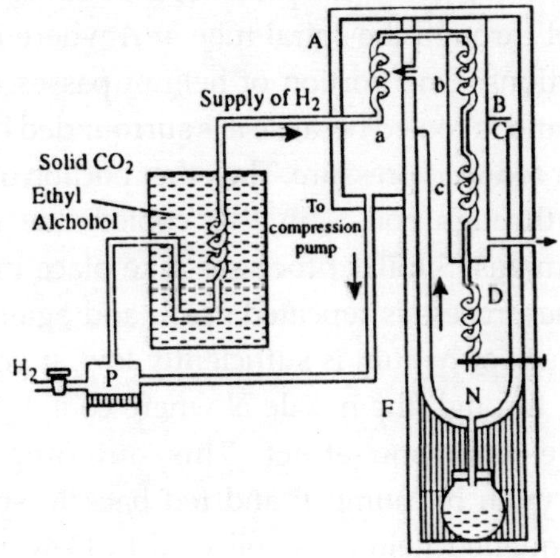

Note : Hydrogen cannot be liquefied by Linde's ordinary apparatus because the temperature of inversion for hydrogen is –83°C. Hence it is essential that H_2 should be cooled first to below –83°C by Joule–Kelvin method and regenerative cooling.

Liquefaction of Helium—K. Onnes Method : The temperature of inversion of helium is –240°C. Kapitza has liquefied helium by precooling it through a coil surrounded by a bath containing liquid hydrogen boiling under reduced pressure. By this process helium was cooled to –258°C. K. Onnes liquefied helium in 1908. His apparatus is shown in fig.

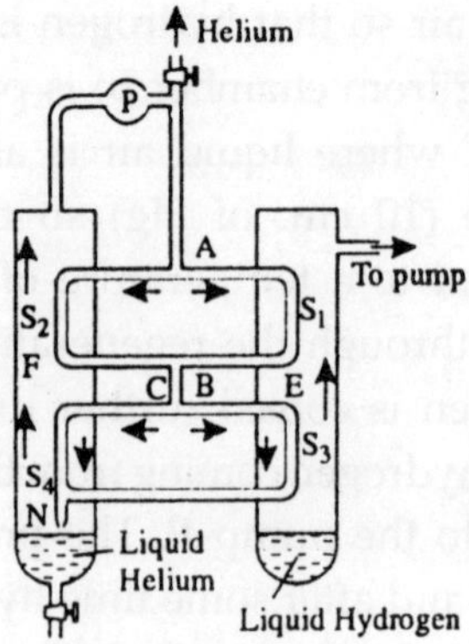

Helium is compressed to a pressure of about 40 atmosphere and is passed through the spiral tube at *A* where it is divided into two portions. One portion of helium passes through the spiral S_1 in *E* and is cooled because it is surrounded by hydrogen boiling under reduced pressure. The other portion of the helium gas passing through coil S_2 in *F* is cooled due to outgoing cooled helium gas. Similar processes take place in the spirals S_3 and S_4. The process is repeated again and again and when the temperature of helium is sufficiently low, it gets liquefied after passing through the nozzle *N* where cooling takes place due to Joule–Thomson effect. This outgoing helium is compressed again by pump P and fed back to spirals S_1, S_2, S_3 and S_4. Liquefied helium is collected in Dewar flask.

Helium I and II : Helium was the last gas to be liquified on account of having the lowest critical temperature (–268°C) of all known gases. It is a colourless transparent very volatile liquid and has the lowest boiling point of 4.2 K at a pressure of 1 atmosphere. It has no normal freezing point and shows quite a few other properties which single it out as almost an exceptional and unique substance.

A peculiarity arising in the helium system is that solid helium can not be obtained merely by lowering the temperature of the liquid. Kamerlingh Onnes failed to solidify helium despite the fact that he reached a temperature of 0.84 K. Solid helium was first obtained by Keeson who subjected liquid helium to very high pressures. The solid is obtained at a pressure of 250 atmosphere at 4.2 K, while only under 23 atmosphere at 1.1 K. Thus, it is necessary to increase the pressure simultaneously while lowering the temperature of the liquid. Later investigations revealed that at a high enough pressure, solid helium could be obtained in equilibrium with the vapour at temperature well above the critical point of the gas. Thus, at 5800 atmosphere, solid helium is obtainable at a temperature as high as 42K. This is the curious property of the helium system that although liquid can not exist above the critical temperature (5.2K), solid can exist if sufficiently great pressure is applied. Hence, at high enough pressures, the melting point exceeds the critical temperature and so the solid helium 'melts' to form the gas.

The phase equilibria of helium are represented diagrammatically in fig. in which all portions are not on the same scale. A study of the phase diagram shows that it is entirely different from that of all other substances. The fusion curve (or the solid liquid phase line) and the saturated vapour pressure curve (or the liquid vapour phase line) do not meet in a point, as in the case of other substances and if we pursue the vapour pressure curve down to lower temperatures it is found that the vapour and the liquid continue in equilibrium down to the absolute zero. Thus the three phases solid, liquid and vapour are never found to coexist or in other words *helium has no triple point* in the conventional sense of terms. This certainly is an extraordinary thing and shows that helium is a unique liquid.

The curve EFK (Fig.) shows the influence of pressure on the melting point. It proceeds upwards to the right, takes a sharp bend at *E* and flattens out at low temperatures meeting the pressure axis at 25 atmosphere indicating that *helium will not solidify even at O K if it is not subjected to pressure exceeding 25 atmosphere.* The S.V. P curve, on the other hand, appears to proceed normally to the left towards the origin ($p = 0, T = 0$) but to the right it terminates at critical point C corresponding to a temperature of 5.2 K and a pressure of 2.25 atmosphere. The point *A* (2.19K) is known as the λ *point* of liquid helium under its own vapour pressure.

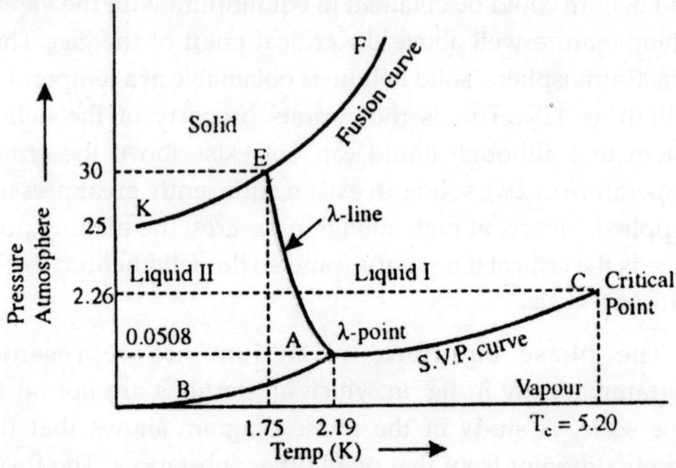

For helium in the liquid phase, there is a phase *transition* called the λ–transition, which divides the liquid state into two phases, helium I and helium II. The fusion curve and S.V. P curve are joined by the λ-line running between the point *E* (T = 175 K, p = 30 atoms) and *A* (T = 2.19 K and p = 0.05 atoms) with helium I to its right and helium II to its left. Thus, helium is present in the liquid form on either side of the λ–line. Kamerlingh Onnes, in the course of his investigations, found that liquid helium shows an extremely interesting behaviour if it is cooled below its boiling point (4.2K) to about 2.18 K. He found that the density passes through an abrupt maximum at 2.19 K decreasing slightly thereafter as shown in fig. The density first rises with the fall of temperature from 4.2 K upto 2.19 K reaches a maximum value of 0.1462 gm/cm^3

at 2.19 K and then decreases with the decrease of temperature. Thus below 2.19 K, the liquid helium which was contracting when cooled now begins to expand.

The specific heat of liquid helium at constant volume has been plotted in fig. The specific heat increases, upto 2.19K and at this temperature there is a sudden and abnormal increase in its value. Beyond 2.19 K the specific heat first decreases and then, increases. The dielectric constant also behaves in a similar manner. The specific heat temperature graph at 2.19 K looks like the Greek letter lambda (λ) and hence this *temperature* (2.19 K) *at which specific heat changes abruptly is called the* λ, *point.* There is a fundamental difference between the nature of the liquid above and below the λ-point (2.19 K). Liquid helium above 2.19 K which behaves in a normal way is called liquid helium I and that existing below this temperature is called the liquid helium II because of its abnormal properties. No heat is involved or absorbed during the transition from one form of helium to another. In other words, no latent heat is involved in the transition He I $\rightarrow$ He II which suggests that:

(a) the entropy is continuous across the curve, *i.e.*, the entropy of He II is practically the same as that of the He I,

(b) there is no change of density during transitions, *i.e.*, the density of both types of liquid is about the same.

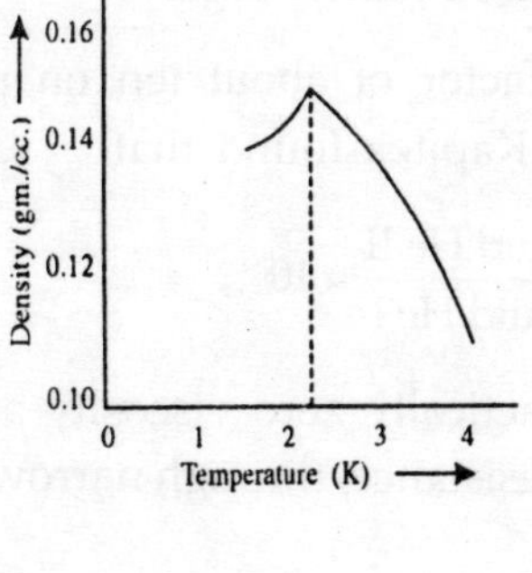

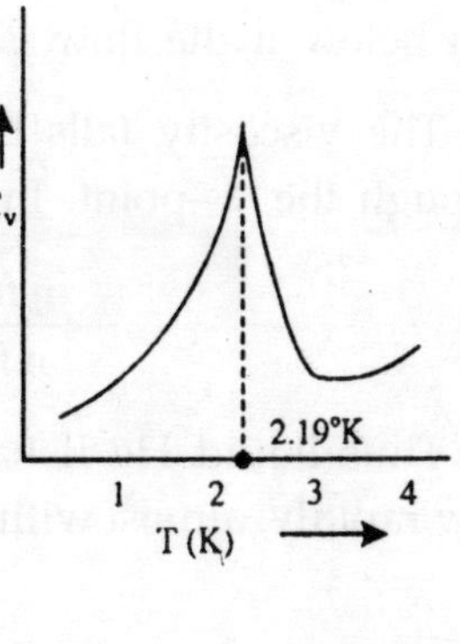

While the viscosity of liquids increases with decrease in temperature, that of liquid helium I decreases. In this respect He I resembles a gas. The viscosity of He II is almost zero and it can flow rapidly through narrow capillaries. Liquid helium I is a normal liquid while helium II presents a very anomalous behaviour in the sense that its thermal conductivity is abnormally high, its internal friction is practically zero and that when it is forced through a capillary the emerging liquid cools while that which remains behind warms up. The densities of the two liquids are about the same, somewhere in the neighbourhood of 0.146 which is far less than that of the lightest of other liquids like ether and gasoline.

Some Peculiar Properties of Helium II : In many of its properties, the behaviour of liquid helium II is quite unlike that of any other substance. It behaves strangely and shows some very unique properties. Some of the more interesting ones are given below

Extreme fluidity: One of the most significant properties of liquid helium II is that its viscosity is abnormally low, even less than that of hydrogen gas and that its rate of flow through capillaries is independent of the pressure difference across them. According to Poiseuille's law, the rate of flow a liquid through a capillary tube or an annular space is inversely proportional to its viscosity. If the rate of flow of liquid helium through a fine annulus is measured as a function of temperature, it is found that at λ–point the rate of flow increases abruptly and below it the flow is extraordinarily large.

The viscosity falls by a factor of about ten on passing through the λ–point. In fact Kapitza found that

$$\frac{\eta \text{ for liquid He II}}{\eta \text{ for liquid He I}} \approx 10^{-3}$$

Thus liquid He II has practically zero viscosity and can flow rapidly, almost without resistance, through narrow tubes.

It is why the liquid He II is called a *superfluid,* a new frictionless state of matter.

The properties associated with it have been characterized as *superfluidity.*

High Heat Conductivity: Heat conductivity of liquid helium I is quite normal, of the order of 10^{-4}, which, in turn, is of the same order as for gases at ordinary temperatures. But He II is found to have an extraordinarily high coefficient of thermal conductivity. The heat transported per unit temperature gradient is several hundred times as great as that in copper at room temperature. He II is said to be about 800 times more conducting than copper and about 13.5×10^6 times more than liquid He I, the absolute value being 820 cals. $cm^{-1} deg^{-1} sec^{-1}$. It is also found that the heat flow is not proportional to the temperature gradient.

Daunt and Mendelson, in 1947, pointed out the striking analogy between the unimpeded flow of helium atoms in liquid helium II and the phenomenon of *super-conductivity.* No trace of electrical resistance is observed in a material which is in a super conducting state *i.e.,* the electrons move freely in the conductor, unimpeded by any electrical resistance.

Formation of Films over Solid Surfaces : Liquid helium can creep along surfaces in the form of highly mobile film generally known as the Rollin (sometimes also called Rollin–Simon) film which is of the order of several hundred atoms thick. The properties of the film were investigated by Daunt and Mendelson who observed the following remarkable behaviour of helium.

If a tube containing helium II be placed in a helium II bath, it is observed that :

(a) If the liquid level inside the tube is lower than that outside, the liquid helium from outside starts creeping into the tube along its surface in the direction of arrows shown in fig. a, the process continues till the levels inside and outside the tube, are equalized.

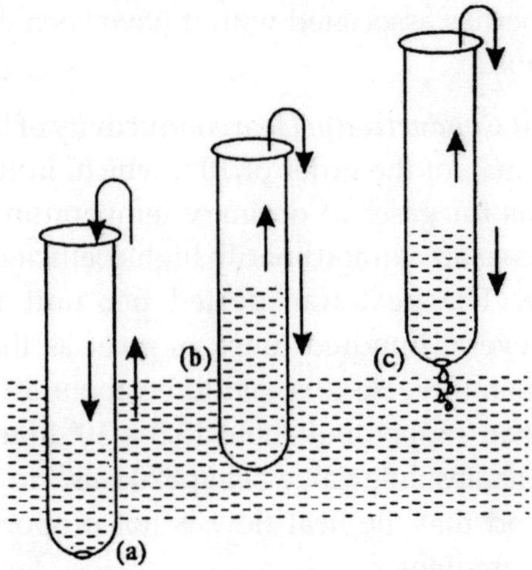

(b) If the tube be raised up so that the level of helium inside it is higher than that outside it, the liquid from inside the tube starts creeping out of it along its surface in the direction of arrows [Fig.b] till the two levels again become the same.

(c) If the tube is lifted entirely out of the He bath, the liquid inside the tube now creeps out along the surface of the tube, collects at its bottom in the form of drops and falls into the liquid below till the whole liquid inside the tube is drained out [Fig. (c)] and the tube is empty.

All these experiments indicate that the liquid helium II seeks the lowest level, but the rate of transfer of the liquid is independent of the difference in level and the nature of the surface. It, however, depends upon the temperature and the perimeter of the surface over which it has to pass.

Thus, liquid He II seems to defy gravity by creeping out of the containing vessel by coating the walls with a thin film of the liquid.

The "Fountain" and "Mechano-Caloric" Effect: Another set of peculiar properties of liquid He II is manifested in the "Fountain" and "Mechano–caloric" effect. The fountain effect, which is illustrated in fig., shows that heat locally supplied to the system produces a pressure difference. The phenomenon was discovered by Allen and Jones in 1938 but the apparatus shown in fig. is one due to Allen and Misener. In the apparatus, He II is taken in a tube *AB* open at both the ends and kept inside a Dewar flask also containing liquid helium II. The upper part *A* of the tube is in the form of a capillary and is kept projecting out of the liquid bath while the lower part *B* with a small opening O at the bottom is packed with fine *emery* powder. The interspaces between the powder particles provide fine capillaries for the liquid He II to flow into the tube. When heat radiations from an ordinary torch are allowed to fall on the powder, liquid He is found to spurt out of the capillary tube in the form of a fountain which has been observed as high as 30 cm. This is a consequence of absorption of energy by the emery powder.

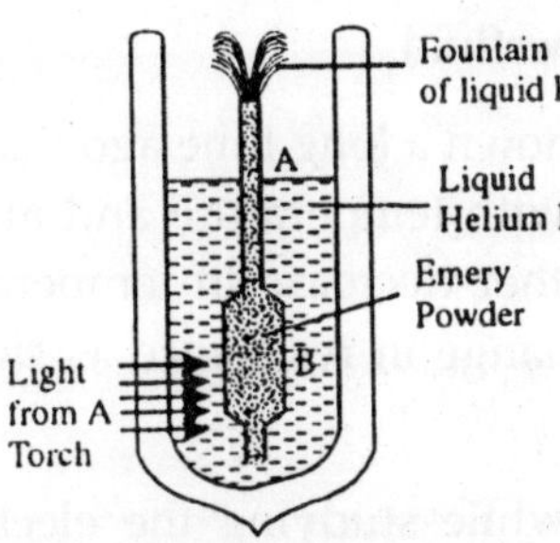

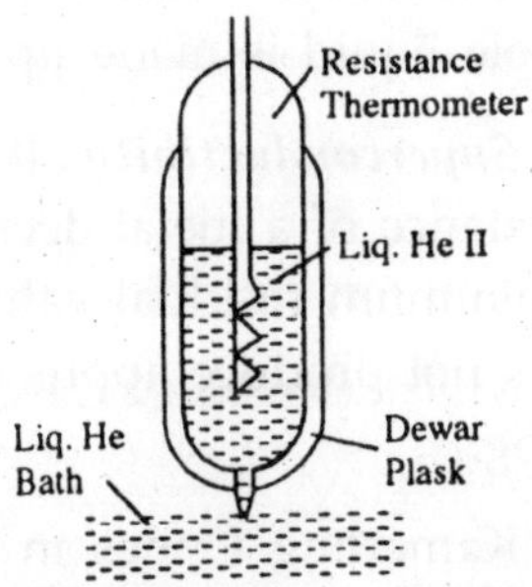

The mechano–caloric effect is illustrated in fig. which is opposite to the fountain effect and shows that the flow of liquid gives rise to a temperature difference. A Dewar flask is filled partly with the liquid helium II and provided with a small hole O at the bottom which is plugged by fine powder. A resistance thermometer is introduced in the Dewar flask and then its top is closed. If now the flask is above the level of the liquid He bath, the liquid from the flask drains out through the powder and its temperature is found to increase by about 0.1°C. If, on the other hand, the Dewar flask is dipped into the bath and the liquid helium is allowed to flow into it through the opening O via the powder, the temperature inside the flask is observed to decrease.

Explanation : *In order to account for the peculiar properties of liquid helium II*, it is generally regarded as a mixture of two different kinds of liquids. One of these is composed of normal atoms with normal viscosity while the other is composed of those atoms which are in the state of lowest energy (zero–point energy) with zero entropy and capable of moving through the normal atoms without friction or viscosity. According to the current views, this part of the liquid consists of 'condensed' atoms and makes its first appearance at λ point. Such a liquid is called a *'superfluid'* and its property to flow through capillary tubes without showing any resistance is called *superfluidity*. On lowering the temperature, the superfluid grows in quantity at the expense of normal component until at absolute zero the whole liquid is made up of superfluid.

Superconductivity : It was known a long time ago that the resistance of a metal decreases with temperature and attains a minimum residual value. Further decrease in temperature does not produce appreciable change in resistance as shown in fig.

Kamerling Onnes in 1911, while studying the electrical resistance of mercury at very low temperatures, observed that the resistance of mercury first decreases regularly like that of any metal but at 4.2 K it *suddenly decreases to zero* (actually becomes about a millionth of its original value at the melting point) as shown in fig. The temperature, at which this phase transition occurs and the resistance suddenly drops to zero, is called the *critical temperature* (T_c). Similar results were obtained by using various other metals like Al, Pb, Sn, In, Zn, Cd etc.

This phenomenon of disappearance of electrical resistance of material below a certain temperature is called superconductivity and the material in this state is called a superconductor.

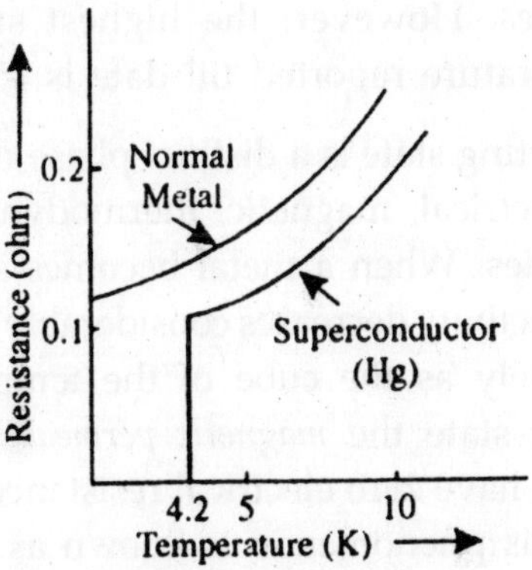

The discovery of superconductivity attracted a large number of workers because the materials with no electrical resistance (negligible heat losses) could be exploited to fabricate powerful and economical devices which consume negligible amount of electrical energy. For example, if a magnet is inserted into or withdrawn out of a coil rapidly, a momentary current flows in the coil but only for the duration when magnetic flux through the coil is changing. If, however, the temperature of the coil is brought down so that if becomes superconducting, then the current will continue to flow for days and even months after the removal of the magnet. Thus, an electromagnet made up of a superconducting material can function for years together even after the removal of the supply voltage. However, due to requirement of very low temperatures, it was not feasible to manufacture such devices. Therefore, a lot of research work was undertaken to develop superconducting materials at high temperatures.

In 1977, a critical temperature (T_C) of 23.2 K was observed in Nb3 Ge (an intermediate compound of niobium and germanium). In 1986, Bednorz and Miller discovered that the superconducting transition temperature in case of La 1.85 Srp

.15 CuO_4 was about 40 K In 1987, a ceramic superconductor of the composition $YBa_2Cu_3O_7$ was discovered which showed T_C equal to 90 K The value of T_c was found to be 125 K for thallium cuprates. However, the highest superconducting transition temperature reported till date is 150 K

Superconducting state is a distinct phase of matter having characteristic electrical, magnetic, thermodynamic and other physical properties. When a metal becomes superconductor, its thermal conductivity decreases considerably and approaches zero approximately as the cube of the temperature. In the superconducting state the *magnetic permeability is also zero.* Superconductors have zero electrical resistance and the *perfect diamagnetism.* This phenomenon is known as *Meissner effect.*

Superconductivity is destroyed by a magnetic field. Even the magnetic field due to the current flowing in the superconductor itself may destroy the superconductivity provided the current is large enough. If, by switching on a magnetic field, the superconducting state is suddenly removed, the metal becomes a good heat conductor. This phenomenon has actually been utilized in magnetic refrigerator.

Bibliography

Adam, A. : *Thermodynamic and Transport Properties,* W.H. Freeman, San Francisco, 1977.

Andrew, L. : *Thermodynamics Physics,* Edward Arnold, London, 1976.

Anotola, S.A. : *Advanced Physics,* Longman, London, 1968.

Anthony, F. : *Statistical and Thermal Physics,* University of California Press, London, 1982.

Atherton, J.C. : *Thermal Physics,* Butterworths, London, 1987.

Bazarov, H. P. : *An Introduction to Thermal Physics,* F.A. Davis Co. Philadelphia, Eisenburg, 1964.

Bemier, G. and Kinet J.M. : *Thermal and Statistical Physics,* C.R.C. Press, Boca Raton, 1981.

Benjamin, W.A. : *Problems in Physics,* Academic Press, New York, 1970.

Bowling, D.J.F. : *Basic Physics,* Chapmann and Hall, London, 1976.

Bray, C. : *Classical Thermodynamics,* Longman, London, 1984.

Canny, M.J. : *Equilibrium Statistical Physics,* Cambridge University Press, New York, 1973.

Cartwright, H.R. : *Elements of Thermal and Statistical Physics,* Oxford University Press, London, 1966.

Cohen, P. : *Control of Enzyme Activity,* Chapman and Hall, London, 1976.

Crafts, A.S. : *Phloem Transport in Plants,* W.H. Freeman, San Francisco, 1973.

Dennis, D.T. and Turpin, D.H. : *Plant Physiology, Biochemistry and Molecular Biology,* Longman, London, 1990.

Douglas, J.S. : *Advanced Guide to Hydroponics,* Pelham Books, London, 1976.

Epstein, E. : *Statistical Physics,* John Wiley and Sons, New York, 1972.

—— : *Thermal Physics : Principles and Perspectives,* John Wiley, New York, 1972.

Esau, K. : *Physical Geology,* John Wiley and Sons, New York, 1965.

Evans, L.T. : *Thermal Analysis of Materials,* W.A. Benjamin, California, 1975.

Forward, D.F. : *Energy and Carbon Metabolism,* Academic Press, New York, 1983.

Ftzler M.E. : *Plant Lectins : Molecular and Biological Aspects,* Academic Press, New York, 1985.

Galliard, T. and Mercer, E.I. : *Recent Advances in the Chemistry and Biochemistry of Plant Lipids,* Academic Press, New York, 1975.

Giese, A.C. : *Cell Physiology,* W.B. Saunders Co., Philadelphia, 1968.

Govindjee : *Thermal and Statistical Physics,* Academic Press, New York, 1975.

Green, J.R. : *An Introduction to Thermal and Statistical Physics,* MacMillan, London, 1911.

Halevy, A.H. : *Handbook of Physics,* C.R.C. Press, Boca Raton, 1985.

Hansen, E. : *Thermodynamics and Transport Properties,* Praeger Publishers, New York, 1966.

Hewitt, E.J. : *Nitrogen Assimilation of Plants,* Academic Press, London, 1982.

Hipkins, M.F. : *Thermodynamics of Chaotic Systems,* Longman Scientific and Technical (ELBS) Essex, London, 1984.

Hopkins, C.R. : *Structure and Function of Cells,* W.B. Saunders Co., London, 1978.

Kirkby, E.A. : *Mathematical Physics,* The University of Leeds, Leeds, 1970.

Loewy, A.G. and Siekevitz, P. : *Cell Structure and Function,* Rinehart and Winston, New York, 1969.

Mazliak, P. : *Lipid Metabolism in Plants,* Chapmann and Hall, London, 1973.

McKee, H.S. : *University Physics,* Oxford University Press, London, 1962.

McNeil, M. Darvill : *Structure and Function of the Primary Cell Walls of Plants,* McGraw, London, 1984.

Meidner, H. and Sheriff, D.W. : *Water and Plants,* Blackie and Sons, Glasgow, 1976.

Mengel, K. and Kirkby, E.A. : *Thermal and Statistical Physics,* Praeger Publishers, New York, 1974.

Mildavin, A. : *Mechanism of Enzyme Action,* Oxford, New York, 1974.

Nomura, M. : *The Control of Ribosome Synthesis,* Edward Arnold, London, 1986.

Novikoff, A. B. : *The Endoplasmic Reticulum : a Cytochemist's View,* Oxford, New York, 1976.

Numa, S. : *Fatty Acid Metabolism and its Regulation,* Oxford, New York, 1984.

Nutman, P.S. : *Symbiotic Nitrogen Fixation in Plants,* Cambridge University Press, Cambridge, 1976.

Ogawa, Y. and King, R.W. : *Plant and Cell Physiology,* Springer-Verlag, Berlin, 1990.

Opik, H. : *The Respiration of Higher Plants,* Edward Arnold Publishers, London, 1980.

Palladin, V.I. : *Physical Metallurgy*, Arihant Publishers, Jaipur, 1988.

Rabinowitch, E. : *Thermodynamics and Statistical Physics*, John Wiley, New York, 1969.

Ramshaw, J.A.M. : *Thermodynamics Physics*, Oxford University Press, New York, 1982.

Rawn, J.D. : *Encyclopaedia of Thermal Physics*, Harper and Row, New York, 1983.

Roberts, J.A. and Tucker, G.A. : *Physics and Chance*, Oxford, New York, 1999.

Ruhland, W. : *Encyclopaedia of Thermodynamics Physics*, Springer-Verlag, Berlin, 1961.

Sabnis, D.D. and Hart, J.W. : *Encyclopaedia of Stastical Physics*, Springer Verlag, Berlin, 1982.

Salisbury, F.B. and Ross, C.W. : *Thermal Physics*, Woodworth Publishing Co., California, 1967.

Siegelman, H.W. and Hind, G. : *Photosynthetic Carbon Assimilation*, Plenum, New York, 1978.

Singer, S.J. and Nicolson, G.C. : *The Fluid Mosaic Model of the Structure of Cell Membranes*, Rinehart and Winston, New York, 1972.

Singh, R.D. : *Cell Biology : A Molecular Approach*, Allyn and Bacon, Boston, 1978.

Slatyer, R.O. : *The Physics of Structural Phase Transitions*, Academic Press, New York, 1967.

—— : *Elements of Thermodynamics Physics*, Academic Press, New York, 1974.

Slavik, B. : *Physics of Crycocrystals*, Springer Verlag, New York, 1974.

Srivastava, H.S. : *Elements of Physics*, Rastogi Publications, Meerut, 1990.

Street, H.E. and Opik, H. : *The Physiology of Flowering Plants*, Edward Arnold, London, 1970.

Thimann, K.V. : *Physics and the Real World,* CRC Press , Boca Raton, Florida, 1980.

Tolbers, P.C. : *Enzyme Kinetics,* Chapman and Hall, London, 1977.

Tolbert, N.E. : *Microbodies—Peroxisomes and Glyoxysomes,* Oxford University Press, London, 1971.

Varner, J.E. and Mitra, R. : *Characterization of Protein Metabolism in Cereal Grains,* Academic Press, New York, 1976.

Verner, J.E. : *Thermal Physics,* Academic Press, New York, 1985.

Vince, Pruce D. : *Elements of Stastical and Thermodynamics Physics,* McGraw Hill, London, 1975.

Walgraif, Daniel : *Introduction to Physics,* McMillan Press, Melbourne, 1969.

Walsh, C. : *Enzymatic Reaction Mechanisms,* Edward Arnold, London, 1979.

Weinheim, D.J. : *Thermodynamics Physics,* Elsevier Biomédical Press, Holland, 1978.

Zeevart, J.A. : *Thermal Plasma and New Materials Technology,* McGraw Hill, London, 1976.

Zhukov, L. : *Thermodynamics of Chaotic Systems,* Springer Verlag, Berlin, 1975.

Ziman, A.L. : *Principles of Thermal Physics,* Worth Publishers, New York, 1982.